Research Funding in Neuroscience
A Profile of The McKnight
Endowment Fund

Research Funding in Neuroscience
A Profile of The McKnight Endowment Fund

By Gabrielle Strobel
Sylvia Lindman, Editor

ELSEVIER

AMSTERDAM • BOSTON • HEIDELBERG • LONDON
NEW YORK • OXFORD • PARIS • SAN DIEGO
SAN FRANCISCO • SINGAPORE • SYDNEY • TOKYO

Academic Press is an imprint of Elsevier

Academic Press is an imprint of Elsevier
30 Corporate Drive, Suite 400, Burlington, MA 01803, USA
525 B Street, Suite 1900, San Diego, California 92101-4495, USA
84 Theobald's Road, London WC1X 8RR, UK

This book is printed on acid-free paper.

Library of Congress Cataloging-in-Publication Data
Application submitted.

British Library Cataloguing-in-Publication Data
A catalogue record for this book is available from the British Library.

ISBN 13: 978-0-12-373645-1
ISBN 10: 0-12-373645-5

For information on all Academic Press publications
visit our Web site at www.books.elsevier.com

Contents

Appendices

Introductions

THE McKNIGHT LEGACY

Erika Binger

It is often when people have the opportunity to interact and connect with others who care about the same things they do that progress is achieved.

In family foundations like ours, philanthropic strategies sometimes evolve from personal interests of individual board members. The McKnight Foundation's neuroscience program is a good example. It has deep roots in our history. My grandmother, Virginia McKnight Binger, and her chief executive, Russell V. Ewald, started the program to honor our founder, my great-grandfather, William L. McKnight. Like many of our programs, our support for neuroscience consequently holds a special place in the hearts of our directors.

William L. McKnight had a keen interest in the human mind, particularly the mechanisms of memory. During the latter years of his life, he spent time, money, and a great deal of effort meeting with physicians, scientists, and others to determine the most promising ways to make a difference in this field. He dreamed of establishing a brain institute to fund research into the puzzle of why memory diminishes as we age. He believed such research could be a great service to humanity.

William McKnight passed away still looking for those answers. After his death, to pay tribute to his passion and interest, the foundation's board of directors explored options for investment in this important work. Following a full year in communication with some of the world's top neuroscientists, the board landed on a structure very similar to what exists today — supporting research by promising individuals, with few strings attached.

This history offers a fuller picture of the impetus behind the program and how it has been modified and improved over the years. Created in 1976, the program made its first awards to neuroscientists in 1977. In 1986, the program acquired its own identity as The McKnight Endowment Fund for Neuroscience, with leading neuroscientists in charge of shaping the program and determining the awards with the oversight of The McKnight Foundation.

For me, most striking is how the endowment fund and McKnight's board work together to make the program as strategically strong and innovative as possible. The neuroscience board has the needed "in-the-field" expertise to autonomously determine the awards' parameters and guidelines. Meanwhile, the McKnight board meets with the endowment fund's board annually and receives and evaluates reports and recommendations between meetings. Over the years, nearly every McKnight director has participated in the endowment fund's annual awardee conference.

At McKnight, we find it truly stimulating (if not always comprehensible) to witness and encourage the wonderful connections the program creates among scientists and across fields, topics, and subjects — helping all to discover new ways to look at and work on their projects. This is a hallmark of many of The McKnight Foundation's programs and one I find especially compelling. It is often when people have the opportunity to interact and connect with others who care about the same things they do that progress is achieved, generally at a significantly greater scale and more quickly than when working in isolation.

Since the 1970s, neuroscience has witnessed many groundbreaking developments. We take pride in knowing that some of the scientists whose work we have been fortunate to encounter have been responsible for some of those breakthroughs. Interestingly, some of this work has even informed our other programs. For example, the endowment fund has supported research on how the brain develops, and knowledge built up as a result of that research lies behind our early childhood program, notably the idea that the infant's brain development is critical to the adult he or she becomes.

We hope you find this report inspirational and informative for your own work in your own community. We are pleased to share what we've learned from the program through the years, and we thank you for your interest.

Erika L. Binger, Chair
The McKnight Foundation
September 2006

A COMMITMENT TO SCIENCE

Carla Shatz

For many, a McKnight award at a crucial time in their careers has enabled a major breakthrough.

In the 1970s, The McKnight Foundation established a program of support for neuroscience in honor of its founder, William L. McKnight. A famously brilliant businessman who had led the 3M Company for nearly 50 years, through its most dynamic period of innovation and expansion, McKnight was concerned about memory loss in himself and his friends as they grew older. He believed support for brain research would be a profound legacy. His successors at The McKnight Foundation, his daughter and son-in-law, Virginia and Jim Binger, and executive vice president Russell V. Ewald agreed.

To pursue this goal, they sought the advice of the late Julius Axelrod, one of the great neuroscientists of the 20th century. A champion of basic science, Axelrod persuaded them to support excellent scientists and allow them the freedom to pursue their own ideas to discover the workings of the brain. Only with that fundamental understanding, he said, would we ultimately be able to conquer diseases of the brain, such as memory loss.

That wisdom has guided McKnight's neuroscience program from the beginning. The program was established as strictly an awards program aimed at fostering excellence, and that is what it remains. McKnight's basic philosophy is to give opportunities to young scientists just starting their careers and to established scientists interested in changing the direction of their research. This philosophy has paid off. Often, those scientists have gone on to open new avenues of inquiry for others. This is the reason McKnight support has influenced the course of neuroscience more than its relatively modest grants would suggest.

Ten years after the foundation first established a neuroscience program, it spun off The McKnight Endowment Fund for Neuroscience as an independent entity, solely funded by The McKnight Foundation. The current structure leaves scientific decisions in the hands of scientific committees, with administrative oversight by the parent foundation. Those of us who steward the program today are still guided by the principles of its founders. Our goal is to support scientists who are willing to work at frontiers to push the field of neuroscience forward, as McKnight has always done.

The McKnight Foundation's constant commitment to neuroscience since the 1970s, a time marked by fluctuations in the federal budget for biomedical research and ups and downs in the emphasis on translational research versus discovery-based fundamental research, is also remarkable. The foundation has been unwavering in its commitment to young investigators and to support for risky visionary research projects. For many, a McKnight Award at a crucial time in their careers has enabled a major breakthrough. Many of these tales of discovery are recounted in this book. McKnight funding for my own lab in 1982 made it possible to embark on experiments that were initially considered too risky for federal support, yet allowed us to discover that the developing fetal brain is bursting with spontaneously generated waves of neural activity as a necessary "rehearsal" for the sensory experience that comes after birth.

As the 21st century began, it was apparent that discoveries over the previous 25 years had revolutionized neuroscience. To take advantage of those discoveries and to keep the program aligned with The McKnight Foundation's initial intent, we designed two new awards, the Technological Innovations in Neuroscience Awards, which support scientists working on new tools to advance the field, and the Neuroscience of Brain Disorders Awards, which support basic research underpinning specific brain diseases.

In 2006, we were proud to celebrate two milestones. It had been 30 years since The McKnight Foundation established a program of support for neuroscience and 20 years since The McKnight Endowment Fund for Neuroscience was created. This book serves as a way of reviewing our story, highlighting accomplishments of our awardees, and reaffirming our commitment to our highest goal: translating basic discoveries to fundamental advances in diagnosing, treating, and preventing neurological diseases. The book also serves as a way for our field of neuroscience to thank the vision and extraordinary generosity of William L. McKnight and his family.

Carla J. Shatz, Ph.D.
President, The McKnight Endowment Fund for Neuroscience
Professor and Chair, Department of Neurobiology, Harvard Medical School
September 2006

Chapter 1

From Donor's Wish to Program Design

HOW McKNIGHT'S NEUROSCIENCE PROGRAM HAS EVOLVED

The McKnight Endowment Fund for Neuroscience bears his name, but William L. McKnight was not a scientist. He is best known as the man who transformed a struggling sandpaper company called Minnesota Mining and Manufacturing into the global powerhouse we know today as 3M. The company is renowned for its breadth and continuity of successful innovation — it owns leading brands in such far-ranging fields as health care, safety and security, transportation, displays and graphics, manufacturing, and home and leisure, and includes such household names as Scotch tape and Post-it Notes. To a large extent, the man who deserves credit for that spirit of diversity and innovation is William L. McKnight. An optimist and a born leader, he launched the company's first laboratory and famously gave his employees the freedom to fail. "Management that is destructively critical when mistakes are made kills initiative," he once said. "And it's essential that we have many people with initiative if we are to continue to grow."

McKnight rose from assistant bookkeeper in 1907 to chairman of the board (1949–1966). Foreseeing his eventual retirement, he cast about in the 1950s for other ways to put his talents to use. He bought several small companies and, in 1953, established a philanthropic foundation, The McKnight Foundation, in Minnesota. The foundation gave mostly to local charities and educational institutions in Minneapolis and St. Paul, until, in 1974, McKnight placed it in the hands of his daughter, Virginia McKnight Binger. He was in his 80s and had long since moved from Minnesota to Florida. His first wife, Maude, had died in 1973.

Research Funding in Neuroscience

William L. McKnight in his office at 3M in the 1940s.

MCKNIGHT'S MOTIVATION

By the early 1970s, McKnight was aware that his memory was failing. One can imagine how a man who had lived by his wits would feel as his best asset, his mind, began to betray him. Not only did he want to do something about it for himself, but he believed brain research could be of great benefit to humanity.

His first major effort in this direction eventually proved misguided. A heart doctor and friend of McKnight's, Edwin Boyle, was interested in whether infusions of oxygen via a hyperbaric chamber would improve a person's memory. Convinced by Boyle that it was possible, and wanting to believe it, McKnight participated in experiments at Miami Heart Institute for 5 years. After all, medicine had nothing else to offer him. Since his days at 3M, he had always been inclined to give science a chance to prove itself, and he underwrote Boyle's research through frequent contributions to the Miami Heart Institute. Ultimately, he reluctantly reached the conclusion that the experiments had little merit. He was right. By 1978, hyperbaric oxygen treatment was shown in randomized clinical trials to have no effect on cognitive impairment in the elderly.

> *It was by no means a given that McKnight's legacy would be used to build the nation's foremost private neuroscience funding program.*

William McKnight's interest in brain science notwithstanding, it was by no means a given that McKnight's legacy would be used to build the nation's foremost private neuroscience funding program. During the early 1970s, the McKnight fortune could well have gone toward funding less distinguished work than it eventually did. McKnight gave serious consideration to using his late wife Maude's estate to start a brain institute in Florida, with Boyle as its leader. Maude had left most of her fortune to expand the existing McKnight Foundation. Even after William and Maude's daughter, Virginia (Ginnie), and Russ Ewald, a former pastor whom she hired as executive vice president, were beginning to run the foundation together, McKnight continued to ponder the merits of a brain institute. Because he was the foundation's honorary chair, it is realistic to assume that Ginnie would have acceded to his wishes, and The McKnight Foundation could have become the channel for her father's plans for a brain institute.

Yet McKnight did not act on this idea. In September 1974, he decided that The McKnight Foundation should receive part of Maude's assets, while the rest would be reserved for the still-nonexistent brain institute. Waiting for the value of 3M stock to go up, he continued to postpone investing. On February 4, 1975, he wrote to Ewald:

> As you probably know, I have been supporting a research study of memory loss in the Miami Heart Institute for some four years and I had planned to broaden this study of the brain backed up by 80 percent of the Maude L. McKnight estate. This whole project has been deferred until the economic condition of the U.S. becomes more clearly known.

CHANGE OF HEART

While McKnight's plans remained in limbo, Virginia's husband, James (Jim) Binger, visited Lewis Thomas, president of Memorial Sloan-Kettering Cancer Center in New York City and author of *Lives of a Cell*. Something Thomas said impressed Jim Binger greatly: "We are on the edge of learning about the brain." In deference to his father-in-law, and optimistic that a foundation program might nurture a scientific breakthrough, Jim Binger proposed that the foundation look for a way to support brain research.

In the fall of 1975, McKnight asked the foundation to consider Boyle's research at the Miami Heart Institute, to which he had contributed $2.5 million over 5 years. The foundation agreed to evaluate the program. Ewald recruited Fred Plum, a renowned neurologist and professor of neurology at Cornell University Medical School in New York City, to lead the evaluation team. Ewald also enlisted Fred King, a professor of neuroscience at the University of Florida, and two Minneapolis physicians, Charles Ray, a spine surgeon, and Charles Burton, a neurosurgeon, to participate in the review. On September 11 and 12, 1975, the team met with McKnight and the Bingers in Florida, talked to Boyle and his researchers, and examined the records and unpublished articles.

Their report was devastating. According to the reviewers, Boyle, a clinical investigator trained in cardiovascular disease and epidemiology, approached the study of age-related memory loss without the necessary rigor, requisite techniques, or formal hypotheses on which to base experiments. Tests placing elderly people with memory loss into individual pressure tanks led to nonspecific effects that had been previously reported by others but were neither interpreted in light of the existing literature nor scientifically followed up. Equipment had been purchased but barely used. The investigators were not familiar with the work of others in the field. Experiments were simplistic; controls were lacking. The technical staff were capable but ran the studies without advice from professional scientists in the pertinent disciplines. The reviewers wrote:

> This appears to be a program without a hypothesis. . . . There was a complete lack of basic neurobiology on aging as it applies to the human or mammalian nervous system. . . . Dr. Boyle himself tends to be diffuse, expansive and unaware of the need to focus on concrete aspects of the specific research problem. It was the unanimous opinion of the Scientific Reviewers that the project did not deserve support from The McKnight Foundation.

Fred King remembered the experience as painful. "Mr. McKnight had made some decisions [to fund the research] he profoundly regretted," King said in a telephone interview in the fall of 2004. "It was one of the more difficult things in my life I've had to do — to tell a man of great faith, Bill McKnight, that he had made a disastrous mistake. He conceded. We recommended that any further consideration of this be done through the foundation and under control of his daughter."

The recommendation was unambiguous. In the end, McKnight decided not to fund a brain institute at all. He had asked Jim Binger to head any institute he might start, but Binger wasn't interested, preferring business to nonprofit administration. "I believe that my turning him down, plus some concern about the validity of the findings of the research and his increasing age — feeling that this is a long, long-term project that he didn't want to get started — all led him to decide he didn't want to undertake the project," Binger said. "He had serious thoughts about it right up to about 1975."

Despite the negative evaluation, McKnight was not willing to abandon Boyle. He wrote to the Bingers in January 1976 to explain that he did not consider the experiments a total loss. "While Dr. Boyle has his faults," he wrote, "I can't downgrade him as much as others have. I suppose I have had more experience in the early years in Industrial Research at the 3M Company than any one else and have lived with many research characters and would rate Dr. Boyle highly among many of them."

He added that the condition of the U.S. economy at that time and his own advancing age had led him to decide against using his late wife's estate to create a brain institute. Instead, he said, "I turned The McKnight Foundation over to Virginia and I wanted the directors to use the income for any purpose they may desire." But, he said, he still wanted something to go toward Dr. Boyle's research, and for that reason he had given $1 million to the Health Sciences Foundation of the Medical Division of the University of South Carolina. Dr. Boyle was moving from Miami to become a clinical professor there, in charge of brain research.

William McKnight withdrew from The McKnight Foundation after relinquishing all funding decisions to his daughter and son-in-law. The Bingers visited Florida and kept him informed, but he never attempted to influence their work. He divested himself of most of his business holdings and died in 1978, at age 90. But he and Boyle maintained a relationship until McKnight's death. Boyle was present at McKnight's autopsy and funeral, as his patron had requested. A few months later, on July 9, 1978, Boyle died in a car accident near his new home in South Carolina.

BIRTH OF A PROGRAM

Following the evaluation of the Miami research, The McKnight Foundation quickly decided to create its own program as a tribute to William L. McKnight.

> *Most advisers favored splitting the anticipated McKnight funds between support for young investigators and a newly envisioned research laboratory.*

Continuing as a consultant to the foundation, Fred Plum tapped some of the best and brightest minds in neuroscience for ideas about how a foundation with relatively modest means could make a difference in the field. Among those he consulted were Donald Tower, director of the National Institute of Neurological Disorders and Stroke; George Cotzias from Brookhaven National Laboratory, the co-developer of L-dopa therapy for Parkinson's disease; Gerald Edelman, Nobel laureate from the Rockefeller University and a man deeply interested in learning and memory; Roland Schneckloth, neurology professor at Cornell University Medical College and chief neurologist at New York Hospital; and Lewis Thomas.

Of these advisers, most favored splitting the anticipated McKnight funds between support for young investigators on the one hand and a newly envisioned research laboratory on the other. The laboratory would integrate basic and clinical memory research between Cornell Medical Center, Memorial Sloan-Kettering, and Rockefeller University. On the face of it, it sounded like a small version of the kind of research center William McKnight had dreamed of, but the foundation eventually decided against the proposal.

In hindsight, it was Schneckloth, the least known of Plum's colleagues, who encapsulated the spirit of the program that would become reality. In a January 1976 memo to Plum, he defined the area where he thought McKnight funds could do the most good:

> Federal support is being directed increasingly toward large program projects, centers, . . . and contracts aimed at solving pressing applied problems. Understandably these large grants . . . are awarded to well established research groups. It has become more and more difficult for the young investigator, following several years of postdoctoral training, to obtain independent support so that he may pursue his own research goals. . . . The potential of the new direction or the unconventional approach is lost at the very time in his life when he is likely to be most original or most productive.

Schneckloth detailed a funding scheme that includes most major characteristics the foundation later adopted for the Scholar Award. This included 3-year, nonrenewable grants to young scholars without overhead payment to the receiving institution, as well as a national, annual competition with selection of awardees made after review by a high-caliber peer group and an interview of applicants.

Putting all this advice together, Plum submitted his recommendation to the foundation in March 1976. He proposed a three-part program:

- McKnight National Scholar-Investigator Awards in clinical brain science and memory research, $25,000 per year to each of five awardees
- An annual, restricted-membership conference on research developments in clinical brain science, including a publication
- A McKnight Laboratory for Brain Science and Memory Research at an existing academic institution

Former presidents of The McKnight Endowment Fund for Neuroscience, from left: Torsten Wiesel, Sam Barondes, Fred Plum, and Corey Goodman.

The foundation formally accepted the recommendation in April, setting aside $750,000 each year for a brain research program. In deference to William McKnight's interest in memory, the directors proclaimed their interest in "stimulating research in neuroscience, especially as it pertains to memory and, ultimately, to a clearer understanding of diseases affecting memory and its biological substrates," according to Virginia Binger's essay in the 1976 annual report. She may have been a little disappointed that the foundation was not specifically trying to cure a disease. One of her dreams was to identify some dread disease and find a cure. But if she was disappointed, there is no indication that she expressed it, except that over the years, she occasionally prodded the scientific directors of the McKnight neuroscience program about whether research progress had advanced enough to tackle a disease.

COMMITMENT TO BASIC RESEARCH

The decision to go with fundamental research was a central one that set the program on its course. It came out of an internal discussion between Ewald and various scientists in the summer of 1976. As a result of the decision, in April, to support brain research, the foundation received two proposals for planning grants to implement the McKnight laboratory—one from Plum and his institution, Cornell, and one from John Shepherd and the Mayo Clinic in Rochester, Minnesota. Ewald's next step proved historic. Rather than simply accept the idea that a focused lab was the way to go, Ewald reached out for an independent opinion. In doing so, he encountered one of the most ardent defenders of basic neuroscience in the country. Ewald approached Julius Axelrod, chief of the section on pharmacology at the National Institute of Mental Health (NIMH). Lewis Thomas had referred Jim Binger to Axelrod, a leader in the field of neuroscience who had won a Nobel Prize in 1970. (For Axelrod's perspective on his role in the development of the program, see the interview with him beginning on page 25.)

> *In reaching out, Ewald encountered one of the most ardent defenders of basic neuroscience in the country—Julius Axelrod.*

Axelrod agreed to help and formed a committee composed of Samuel Barondes, professor of psychiatry, University of California, San Diego School of Medicine; Edward Evarts, chief of the laboratory of neurophysiology, NIMH; Seymour Kety, director of the psychiatric research laboratory, Massachusetts General Hospital; and James McGaugh, vice chancellor of academic affairs and professor of psychobiology, University of California, Irvine.

The group met on July 14, 1976. In addition to the committee members, those in attendance were Plum; Ginnie and Jim Binger and their daughter Cynthia Boynton, a foundation board member; and the foundation's two staff members,

Russ Ewald and Marilyn Pidany. They decided to concentrate on the awards program initially, drawing up an announcement and compiling a list of scientists to be invited to apply. In September the committee met again and appointed awards review committees. The committee also asked Ewald to pursue an annual or biennial conference on the state of the art in neuroscience, for 1977 or 1978. After this meeting, the committee disbanded.

If the awards and the conference program took off, why did the third part of Plum's proposal not move forward? In essence, the foundation decided to go with the vision of basic research. In his proposal, Plum noted what had been the foundation's earliest intentions:

> Several investigators are doing good fundamental research on how the nervous system forms memory. Here one can cite the work of McGaugh at Irvine, Barondes at San Diego. . . . In all instances, however, the laboratories are presently directing their efforts at basic work of the most abstract kind involving exclusively lower animals or test tube efforts. None of them are multidisciplinary in their efforts, and none of them presently believe that anything in their work justifies carrying the laboratory efforts to application in patients. My understanding is that this basic a program, and one not at all directed at diseases of memory in man, is not what the Foundation desires.

For his part, Axelrod believed that basic research was more likely to lead to new therapies than targeted clinical research. Axelrod concurred with the conclusion of an unusual paper *Science* magazine published in April 1976. He sent the paper to Ewald before the ad hoc committee was to meet in July to decide the fate of the proposed McKnight Laboratory, and it may well have influenced Ewald.

To Axelrod, excellence, not clinical relevance, was the key to success.

In the article, Julius Comroe, a distinguished professor of physiology at the University of California, San Francisco, and Robert Dripps, a professor of anesthesia at the University of Pennsylvania, addressed the question of whether it was targeted, disease-oriented research or basic research that had made the larger contribution to the top 10 clinical advances in cardiovascular medicine during the past 30 years, open-heart surgery being one example. The authors led a team that spent several years on an analysis of thousands of scientific papers published in the past 100 years that had built the knowledge base necessary for those 10 medical advances. Perhaps a strange pursuit for medical professors, the study was an attempt to move an ongoing national debate about the relative merits of basic versus "mission-oriented" research past the anecdotal conversation that predominated at the time. The debate was crucial for the future direction of national science policy, and Comroe and Dripps hoped that their data would put such decisions on an objective footing. In any event, they found that basic, undirected research had made the overwhelming contribution to those medical breakthroughs, and their finding resonated with Axelrod. To him, excellence, not clinical relevance, was the key to success.

Finally, Axelrod believed that talented researchers across the country were the field's greatest asset and that money should go directly to them, not to an institution. In an interview a few months before his death in December 2004, at 92, Axelrod explained his reasoning in detail. "We wanted basic information about how the nervous system retained information," he said. "In neuroscience that was radical at the time, but we thought it was doable. Above all, we sought imagination and great ability in the awardees."

Axelrod's arguments persuaded Ewald and the Bingers, and they went forward accordingly. In 1977, Jim Binger wrote to William McKnight to explain how the program was set up:

> There is probably a better way to do this, but I haven't thought of it. I'm impressed by how very little is understood about the functioning of the brain. This work is very far out and needs a lot more successful basic research done before meaningful applied work can be undertaken.

FOUR CORE PRINCIPLES

The program began with two types of award, given first in 1977. One, the McKnight Scholars (later Scholar) Award, was meant to nurture young investigators just starting their own laboratories. Young scientists at this stage are poised to embark on perhaps the most creative phase of their careers but for the most part do not yet have access to federal funding, which in the case of neuroscience comes mostly from the National Institutes of Health (NIH). Scholars would receive $25,000 for 3 years. Plum headed the founding selection committee, which included Ann Graybiel of the Massachusetts Institute of Technology; Nelson Goldberg of the University of Minnesota; Solomon Snyder of Johns Hopkins University; Barondes; and McGaugh.

The other award, the McKnight Award for Research Projects, later renamed the McKnight Senior Investigator Award, was to give distinguished professors the freedom to diverge from mature areas of research and open up new, high-risk lines of investigation or to bring new techniques and perspectives to their laboratories. They were to receive up to $100,000 per year for a flexible number of years. Axelrod oversaw the selection process for this award, serving with Murray Jarvik of the UCLA School of Medicine, who later made key discoveries about nicotine addiction; Oliver Lowry of Washington University, who had designed widely used methods to measure proteins in solution; Eugene Roberts of City of Hope Beckman Research Institute, who had identified GABA, a major inhibitory neurotransmitter in the brain and spinal cord; Evarts; and Kety.

Four principles shaped the program in these early years:

1. *Ewald's personal integrity imbued the program with an ambitious tone and purpose.* Even though Ewald oversaw numerous other philanthropic activities at The McKnight Foundation, he took a particular interest in its fledgling brain research program. He made sure to meet all Scholar and Senior awardees of the first funding round at a special inaugural dinner, and in 1978 he visited them each in their laboratories. In the beginning years, Ewald worked closely with Axelrod and Evarts as they refined the selection process and experimented with the award cycles and distribution of funds between the awards. Through the 1980s, Ewald worked closely with the committees.

2. *Axelrod's determination that scientific merit and peer review take precedence over academic pedigree and connections shaped the award selection process.* Barondes, who shared this view, recalled in an interview in 2005:

> We aimed for excellence from the start. We knew that an outstanding committee would attract and select outstanding applicants. And once a very high standard was set, the word would get out that this was an award to be proud of. So I and others on the original Scholars Award Committee worked hard to establish this precedent. If we did not have sufficient outstanding applicants to claim all the awards we chose to select only the top ones, rather than compromise on quality—and reserve unfilled awards for another year.

3. *The founding scientists placed their bets on broad, basic science.* They considered the foundation's ultimate goal of helping with memory disorders beyond immediate reach in the late 1970s. In their estimation, scientists' grasp of the cellular and molecular mechanisms of brain function, as well as the brain's circuits and systems, was in its infancy and plainly inadequate to provide a sound foundation for directed research.

The brain's workings—the plasticity of its synapses, the firing of its circuits, their wiring during development, their responses during learning—were still too much of a mystery to tackle its malfunctioning in a rigorous way. Axelrod and his fellow scientists believed that the best strategy toward realizing the foundation's interest was to combine excellence and breadth. The selection committees would work to identify the most talented and energetic scientists, but then give them free rein to pursue creative questions of fundamental brain research. In time, this approach would generate insight into the basic underpinnings of memory, from which would grow an understanding of disease and, in turn, the ability to develop mechanism-based therapies.

This third principle was the subject of some internal debate. Some scientists involved early on in the McKnight neuroscience program, notably Robert Terry, James McGaugh, and Fred Plum, believed that disease-oriented research also deserved funds at the outset. Yet the focus on basic science prevailed in 1976 and indeed through the 1990s, when it was revisited.

4. *The McKnight Foundation decided to invest in people more than in specific proposals.* In practice, this meant that the award money came with

few restrictions. Indeed, the only restriction was that the funds not be spent on indirect (i.e., university) costs. The awardees would be allowed to follow their instincts and even change the stated research project if preliminary data or a tantalizing new idea pointed them in a different direction. This philosophy of trust in a rigorously selected scientist set The McKnight Foundation apart from other funders, notably governmental ones, which expect that funds be spent exclusively on the submitted proposal. To make sure awardees were indeed the best among their peers in any given year, the selection process for the Scholar Award was weighted less toward particular details of an applicant's research proposal and more toward its originality and conceptual clarity. The impression finalists gave during a personal interview with the selection committee played an important role as well.

To this original program, the foundation in 1981 added a third award, completing Axelrod's vision of a three-tiered award program. It targeted scientists who fell between the Scholar and Senior awardees in that they had already made their mark in neuroscience but were still short of tenured professor positions. This award was established as the McKnight Development Award and later renamed the McKnight Investigator Award. The Senior Awards Committee assumed selection of candidates for this award under its aegis.

Finally, in 1983, the foundation authorized a fourth award, called the Director's Award. Its purpose was to serve as a ready pool of discretionary funding with which the neuroscience program could react promptly to major breaking developments in neuroscience pertaining to memory and its disorders even while its funds were committed to 3- to 8-year projects at any given time. (In the first years of the program, Senior Awards were given for 3 years, then evaluated by the award committee for a 5-year continuation.) Early advances in the molecular biology of Alzheimer's disease had inspired Axelrod and Evarts to give such a nimble funding mechanism a try. Cognitive neurologist Marek-Marsel Mesulam, then of Beth Israel Hospital in Boston, and Nobel laureate James Watson, who had co-discovered the double helix structure of DNA, received this award, yet it did not develop a distinct identity next to the Scholar, Investigator, and Senior Awards, and the endowment fund used it infrequently.

The McKnight neuroscience program was a pioneer in the 1970s funding environment.

The McKnight neuroscience program was a pioneer in the 1970s funding environment. Besides NIH and the National Science Foundation (NSF), few funding resources existed for neuroscientists at the time. NIH grants were hard to come by for newly emerging investigators who did not have the prior data required to substantiate an NIH proposal. NIH and NSF funds tended to support the continuation of existing research rather than risky ideas that can lead to paradigm shifts. Established in 1934, the Sloan Foundation was one of the few established private philanthropies that funded basic research in science and engineering, but its interests were not specific to neuroscience. Other existing philanthropies tended to fund applied research, focus on specific diseases, or support institutions rather than individual scientists.

An exception was the Markle Scholar Program by the Markle Foundation. It had inspired Schneckloth's vision of a funding program when Plum had asked his advice in January 1976. According to the online Rockefeller Archive Center, the Markle program had since 1947 supported academic medicine and research from 90 medical schools around the United States and Canada but ended in 1969 when the foundation decided to pursue interests in mass communications instead. The Markle fellowship had funded young faculty with the stated intention of keeping them in academic medicine, where they could achieve something new and important. The program had produced leaders in academic medicine at medical schools and in government, such as Robert Marston, who directed the NIH during the Nixon administration, and Alexander Schmidt, commissioner of the Food and Drug Administration from 1973 to 1976. Being able to call oneself a Markle Scholar was a badge of distinction and prestige for medical researchers at the time, much as being a McKnight Scholar is for neuroscientists today.

In addition to the research awards, The McKnight Foundation in 1976 began exploring the idea of a conference for awardees, a tool the Markle Scholar program had used effectively to create a network of peers. The foundation and its scientific advisers believed that providing a forum for awardees to meet and exchange ideas and data could be a powerful means of creating momentum for first-rate neuroscience. A conference would foster a sense of community among McKnight-funded scientists and help keep them involved past the expiration of their grants. The conference therefore could be a platform linking the pillars of the program, that is, the research awards.

The first conference was held over 3 days in February 1980 at Spring Hill Conference Center in Wayzata, Minnesota. Axelrod gave the opening lecture. Through the 1980s, awardees past and present, the members of the selection committees, and a few special guests met every other year at Spring Hill with Ewald and the foundation directors and staff. Starting in 1990, the foundation experimented with different locations. Currently held over 3 days in June at the Aspen Institute in Colorado, the conference has become popular among awardees and committee members. In 2000, it became an annual event.

A CRITICAL JUNCTURE

Ewald's efforts to leave behind an independent charitable organization have created an acclaimed brand name for neuroscience research funding.

The neuroscience funding program continued in this way until the mid-1980s, with three awards and biennial conferences in Minnesota and the boards reporting to The McKnight Foundation. Russ Ewald oversaw the program with the steady support of Ginnie and Jim Binger and the board of The McKnight Foundation. As the first generational change in the program's stewardship was drawing near, Ewald was concerned that a future McKnight board might not want to spend so much time overseeing all the neuroscience program activities, potentially jeopardizing the very existence of the program. He wanted to ensure a continued commitment to neuroscience after he and Virginia retired. With Ewald and Jim and Ginnie Binger backing the idea, the foundation board in 1986 set up a new organization that became The McKnight Endowment Fund for Neuroscience (EFN). The EFN would have its own board of directors to approve awardee selections, freeing the parent foundation from the need to ratify the slates recommended by the committees.

The initial plan was to start the EFN with a $20 million endowment, but the foundation discovered that such a payment would not be a qualifying distribution for a not-for-profit philanthropic organization because its recipient was too close to the parent foundation's existing operations. After taxes, the remaining endowment would yield a smaller annual budget to disburse for research funding than the neuroscience program had received from the foundation until then. The McKnight Foundation decided instead to manage the $20 million in an earmarked account and fund the EFN with annual grants from its proceeds. The new entity would still be called an "endowment fund," even though it would not have its own endowment. And it would still have an autonomous board and officers, but with representatives from the parent foundation acting as liaisons.

The establishment of the EFN made way for changes in the roles of some of the founding scientists. Fred Plum, who had contributed immensely to the program and was highly regarded for his accomplishments in coma research and for his diagnostic skills as a neurologist, became president of the EFN board. In turn, he relinquished the chairmanship of the Scholars Committee, which he had led since its inception in 1977, to Sam Barondes, who had served on that committee from the start.

The newly formed EFN was managed initially in Plum's office at Weill Medical College of Cornell University in New York. But the administration of the various programs and frequent interactions with applicants, grantees, and committees proved too demanding for an academic office. In 1989, The McKnight Foundation reclaimed the managerial role, drawing up an agreement whereby the foundation provides fund management and administrative services to the EFN, while the scientific directors run the awards programs. Marilyn Pidany oversaw the program until she retired in 1999, when the responsibility transferred to Kathleen Rysted, who had already been working with Pidany on the program for a few years. The EFN has operated efficiently in this way ever since; indeed, the administrative backing of a major foundation has been one reason why the EFN has achieved a disproportionate impact since the 1980s.

The scientists always appreciated the management support. When Ewald in 1985 discussed his intention to make the program independent, the scientists did not jump at the chance to "own" the program but said they would prefer it to remain within The McKnight Foundation. The present arrangement has helped the program thrive because the EFN directors have enjoyed the freedom to focus on what they do best — select talented scientists and take charge of the program's conceptual direction — without being burdened by management requirements.

Ewald's efforts to leave behind an independent charitable organization have created an acclaimed brand name for neuroscience research funding that continues to enjoy the backing of McKnight's board. Virginia's daughter Cynthia Boynton, who succeeded her as president, continued her mother's enthusiastic support for the neuroscience research program, as did her brother, James (Mac) Binger, and the foundation's current leadership, which to this day consists largely of William and Maude McKnight's descendants and their spouses.

In 1999, the EFN reshaped the program in a way that incorporated the foundation's original wishes and reflected enormous progress in the field over the previous 20 years.

How has the system of annual grants served the EFN, and how much money has it been able to award over the years? Initially, The McKnight Foundation agreed informally to make available annual grants of about $1.75 million for 15 years. In 1994, with inflation steadily eroding the EFN's ability to fund research and with a positive outside evaluation attesting to the program's value, Barondes met with the board of directors of The McKnight Foundation and requested a multiyear commitment that would enable him to plan the EFN budget proactively, move funds among the awards to respond to changes in the applicant pools, and provide the means for discretionary awards and new initiatives. Barondes also asked for a general increase in funds to expand existing EFN programs. In response, The McKnight Foundation committed $10 million from 1995 through 1998, contingent on reevaluation of the fund's award programs.

In 1999, the scientific directors reshaped the EFN program in a way that incorporated the foundation's original wishes. Enormous progress had occurred across the field of basic neuroscience in the past 20 years, in part catalyzed by McKnight grants to hundreds of outstanding investigators. In a carefully reasoned 10-year funding proposal submitted that year, the directors laid out how they would ensure that the EFN would remain in a position to leverage a disproportionate benefit from its comparatively small award payments. The proposal satisfied the foundation's directors not only that their money had been well spent up to then, but also that the future direction the scientists outlined was worth further investment. The foundation decided to set aside annual grants of

$4.4 million until 2010, guaranteeing a continued funding stream for a third decade.

LESSONS LEARNED

The 1990s were marked by several adjustments in the fund's governance and strategic planning even while it continued to build a stellar record of fueling important discoveries in neuroscience. To the degree that these adjustments offer lessons for philanthropy in general, this history will tell them here.

To begin with, the first generational change occurred on the EFN board of directors, in 1989. In April of that year, Ewald resigned as executive vice president of The McKnight Foundation, a position that had guaranteed him a seat on the board of directors of the EFN. The fund being his pride and joy, Ewald would have liked to stay on its board after leaving the foundation. But that did not happen, and his successor, Michael O'Keefe, took the seat on the board reserved for the foundation's chief executive. Also in 1989, Axelrod, Kety, and Plum retired to make room for young scientists with fresh ideas. Plum, who had helped design the original program and had overseen the Scholar Award well, left reluctantly.

Since then, the EFN has established a policy to accommodate outgoing directors who wish to keep contributing. They have assumed emeritus, nonvoting positions on the board and have at times taken on special assignments.

The newly reconstituted board had new ideas, particularly about how to identify research to support. The first round of Senior Awards in large part had gone to some of the scientists who were shaping the McKnight program or who served on the Scholar Award Committee or to some of their close colleagues in New York.[1] At that time, the program was still finding its way. General sensitivity about conflict of interest in research was less acute than now, partly because fewer academic scientists had industry ties and partly because philanthropy was smaller and less prominently in the public eye. Conflict-of-interest policies for scientists were not yet in wide use. But viewed in retrospect from the year 1989, when a new generation of leaders came on board, awards to scientists who also represented the McKnight program suggested the time had come.

After 1990, no scientist serving on any selection committee or on the board received funds for his or her own lab while representing the foundation. More broadly, starting with the Scholar Award, the EFN also adopted rules designed to avoid favoritism during the selection process. In 1993, Corey Goodman formalized rules for the Scholar Committee whereby committee members with

---

[1] Julius Axelrod, Edward Evarts, and Seymour Kety never received McKnight Awards despite serving for more than a decade.

a conflict of interest (either because they are from the same institution as an applicant or have mentored the applicant) do not ask questions of that applicant during the applicant's interview. Moreover, initial votes were taken based on the application before the board discussed the applicant, and an Olympic-style scoring system was used to determine the top six candidates. In 2001, the EFN extended such policies to all three committees. Committee members agreed to leave the room during voting on applicants from their own institution and to abstain from voting on applications by former postdoctoral fellows or trainees. In addition, EFN board members now complete an annual conflict-of-interest statement that bars them from deriving personal gain from their association with the EFN and discloses relevant ties to other organizations.

A different tool for avoiding ingrained practices and easing the retirement of future EFN scientists came in the form of term limits. The board in 1994 established terms for directors, officers, and researchers serving on award selection committees. Rotation schedules were set up to balance infusion of new blood with continuity. Initially, the term limits were written into the bylaws of the EFN, but later the board decided to handle them through guidelines to allow for flexibility in making changes as needed.

ACCESSIBILITY AND OPENNESS

Over time, as the McKnight name spread, the awards became coveted and prestigious and the closed nomination mechanism turned into a liability.

Another issue the directors wrestled with during the 1990s concerned the mechanism of inviting proposals for some of the awards. When the founders established the program, they were wary of being inundated with a flood of applications if all awards were to be announced openly. For the Scholar Award, the inaugural selection committee decided to draw up application forms and to publish an announcement in the *Journal of Neuroscience*. The Senior Award and the Investigator Award were handled differently. Their committee decided to solicit applications by invitation only. Indeed, in 1976, the committee members initially decided not even to draw up an application form but simply to tell the candidates they had preselected — about twice as many as would eventually be chosen — what information to provide.

The founders believed that an accomplished board would know the field well enough to be able to identify the best researchers. Essentially, the approach seemed efficient, and in the early years it worked well. Among the senior scientists chosen in the 1970s and early 1980s who have remained highly productive leaders afterward are Paul Greengard, Eric Kandel, Solomon Snyder, and Chuck Stevens. Early McKnight Investigator Award recipients Eric Knudsen, Anne Young, Paul Patterson, and Larry Swanson are examples of midcareer scientists who have become highly successful since then, to say nothing of the awardees who later became EFN board members, such as Corey Goodman, Thomas Jessell, and Carla Shatz. In later years, too, the award committees identified the best and the brightest of neuroscience.

Yet over time, as the McKnight name spread among the wider neuroscience community, the awards became coveted and prestigious and the closed nomination mechanism turned into a liability. It created a perception that McKnight Awards were being held among an exclusive network of friends.

It was primarily the Senior Investigator Award that generated this impression, and its feature of offering award extensions compounded the problem. During its second funding round, in 1980, all six Senior Awards were extensions of previous awards originally given in 1977. Four of the six recipients were from New York City and a fifth, Solomon Snyder, had spent his postdoctoral years in Axelrod's lab and had then become an outstanding neuroscientist in his own right. Only one new name, Alzheimer pathologist Robert Terry, joined the original group in the years between 1977 and 1985. In 1985, the circle of Senior awardees widened with 10 fresh appointments but then narrowed again when six of them tied up a large part of the 1988 budget with extensions. Thus, the McKnight funds for Senior Awards were distributed among a fairly small number of scientists during these years. By contrast, the Scholar and Investigator awards did not grant extensions.

In 1979, Axelrod and his fellow advisers had deliberately set up this 8-year funding period with the intent of giving long-term support to a select group of particularly productive basic researchers. By 1989, The McKnight Foundation's new executive vice president, O'Keefe, began to question whether the distribution of funds going to young versus seasoned investigators struck the right balance. Related questions were whether the slate of candidates for the Investigator and Senior awards was broad enough, whether award extensions should continue, and whether the occasional practice of handing out discretionary awards outside of the Scholar, Investigator, Senior, and Directors' awards ought to be formalized.

In managing his predecessor's program, O'Keefe challenged the directors to keep on their toes and periodically refresh their operations. Some EFN directors hardly needed prodding. Goodman began to experiment with the application process right after he took over as chair of the Scholars Award selection committee. Then again, this award had never drawn much criticism. It most closely represented the foundation's wish to give wings to talented investigators during their early years when they are at their most dynamic and least financially secure. Besides, it was arguably the most competitive of the three awards in that many hundred initial letters of intent came in every year, generating 8 to 13 full applications per granted award, depending on the year. For the Investigator and Senior awards, the ratio of applications to grants hovered between 1.8 and 2.3 because in the early years, the committee invited only a few more people to apply than would eventually receive awards.

The stewards of the Investigator and Senior awards had genuine disagreements with O'Keefe. He questioned whether financing established scientists was a good use of the foundation's funds. Both awards were highly successful in that they had built an impeccable record of picking accomplished scientists. Many of the pilot projects begun with McKnight funds later flourished into mainstream projects subsequently funded with larger grants from the NIH. As a group, these scientists tended to enjoy adequate research means. But even for them, the NIH funding rate did not exceed 25 percent at the best of times. They had to divert committed funds to support risky studies for which no prior data existed as yet. In short, established scientists, too, wanted access to unrestricted money for their wilder ideas. That they deserved McKnight funds was a deeply held view put forth by spirited, charismatic leaders, and consequently change came more slowly. But come it did. The last two Senior Award extensions were granted in 1994, and the Senior and Investigator awards themselves were nearing the end of their run.

OUTSIDE EVALUATION

Concurrently with these discussions, O'Keefe requested an evaluation of the neuroscience program by an independent consulting firm. O'Keefe, a seasoned administrator, believed in close oversight and thought it was time for a thorough review of the program. Indeed, he had conducted reviews of several of the foundation's other programs as well. The purpose of the evaluation was to inform the McKnight board of the value and performance of their brain research investment and to guide the EFN directors in their strategic planning.

> *The report invited the directors to come up with new ways of integrating basic neuroscience with studies of brain disease.*

To conduct this review, he engaged Abt Associates of Cambridge, Massachusetts. Abt sent investigators to meet with Barondes and other EFN board members, examined the records of all the awards that had been made, and analyzed the productivity of the scientists who had received awards. In 1994, Abt Associates handed in a positive report that endorsed the program's scientific record as first-rate and found no major shortcomings in its administration. Its suggestions ended internal debate on several issues and provided a blueprint for change.

First, the selection committee for the Investigator and Senior awards would no longer handpick candidates but would invite nominations from the chairs of relevant academic departments all across the country. Second, the report strengthened a shift the EFN had previously begun, whereby it added a greater proportion of systems-based studies to its present docket of predominantly cellular and molecular approaches. This change was taking advantage of the recent development of new techniques for studying brain function and cognition at these levels of analysis. Third, the report triggered the establishment of term limits, a change the board had been considering for some time. Fourth, the report invited the directors to come up with ways of integrating basic neuroscience with studies of brain disease.

In this way the report stimulated a thinking process about how the neuroscience landscape had changed in the past 20 years and what sorts of changes would best position the EFN program to ensure it would continue to have a high impact in the future. For a while, the board considered establishing new programs, such as a McKnight Journalist Program or a McKnight Prize in Learning and Memory, but eventually it decided to stay focused on finding and supporting the best researchers. Consequently, future changes would focus on the award programs.

In 1995, Eric Kandel's term as chair of the Senior/Investigator awards review committee ended, and he passed the baton to Gerald Fischbach, who at the time headed the department of neurobiology at Harvard Medical School. Together with two new members, Richard Scheller of Stanford University and Stephen Heinemann of the Salk Institute for Biological Studies, the committee's face changed considerably that year. In 1997, these scientists used an open nomination process to select Senior awardees one last time to receive funds through 1999. After 20 years of helping accomplished scientists to branch out into new research areas, the award had run its course.

A New Era

The Technological Innovations Award grew out of the recognition that researchers in chemistry, physics, and engineering were inventing technologies that should be brought to bear on neuroscience.

Sam Barondes's stewardship of the EFN reached its term limit in December 1998. He had served as president for 9 years. Torsten Wiesel, a Nobel laureate who had retired as president of Rockefeller University the previous fall, succeeded Barondes for 2 years. Wiesel represented continuity more than change. During his presidency, younger directors, led by Corey Goodman and Carla Shatz, both then at the University of California at Berkeley, and Tom Jessell, of Columbia University, became the de facto leaders in planning for the next decade.

With Barondes, who would remain on the board for several more years, they formed a planning group that worked to overhaul the program. The overhaul not only rethought the program in light of advances in neuroscience since the 1970s but also addressed growing concerns among The McKnight Foundation board that the EFN was not making enough progress toward solving the problems of memory loss. The goals of the McKnight family remained organized around alleviating suffering, and the 20-year point in the program's history offered an opportunity to look at new ways of addressing those goals.

As their first major initiative, the planning group replaced the Senior Investigator Award with a new award conceived by Shatz—the McKnight Technological Innovations in Neuroscience Award. It grew out of the recognition that researchers in chemistry, physics, and engineering were inventing technologies that should be brought to bear on neuroscience.

The field of neuroscience had matured considerably since the inception of the McKnight program, but it urgently needed new techniques to visualize and analyze neuronal function. While there is general consensus that importing new technologies speeds up progress in any given field of biological research, it is also true that the communication across scientific languages and the close interaction necessary for such a welding of the minds are difficult to sustain in practice. These goals need a catalyst. Pushing the interface of neuroscience and other disciplines therefore represents a niche where a McKnight Award could make a difference. It also coincides with The McKnight Foundation's interests in that this kind of research is early-stage, high-risk work that requires seed funding and falls outside the purview of other funders.

Every year starting in 1999, the award would offer $200,000 over 2 years to individual scientists or groups of collaborators to develop new imaging, sensor, and other technologies to explore neuronal activity. Lubert Stryer of Stanford University, who co-invented the DNA microarray, agreed to oversee award selection. The initial announcement for this new award drew 130 proposals, a surprisingly high number that bespoke both the funding need and the recognition of the McKnight name.

> *Both the opportunities and the challenges of translational research had changed to where the directors saw a new rationale for a McKnight Award.*

The second major change occurred around the fusion of basic and applied neuroscience. The board felt that much as its predecessors had been wise to defer translational research in 1976, the intervening years had witnessed such remarkable progress on the basic research front that the time had come to revisit the issue. In part thanks to the basic neuroscience advances facilitated by McKnight, efforts to apply such insights to the diagnosis, prevention, and treatment of brain disorders in 1999 could draw on a wealth of tools and knowledge and consequently deserved support. Both the opportunities and the challenges of translational research had changed to where the directors saw a new rationale for a McKnight Award.

For example, M.D.–Ph.D. physician investigators are ideally suited for translational research but frequently find their research aspirations choked by the clinical requirements at their institutions. Basic scientists need help to overcome perceived barriers to disease-related research, and medical doctors need a period of freedom to master the methods and standards of investigation of basic neuroscience. While the NIH and special patient interest groups have traditionally supported clinicians who focus on the basic science of a particular disease, Ph.D. researchers willing to venture into the study of a disease have nowhere to turn for funding. But equipped as they are with modern tools of genetics, molecular biology, and functional imaging, it is they who could make the greatest advances in understanding a disorder.

To entice the very best basic neuroscientists to turn their attention to the interface of basic and clinical neuroscience, the EFN created the McKnight Neuroscience of Brain Disorders Award. Originally Barondes's idea, the award continued the McKnight philosophy of breadth, as it supports translational projects not only in diseases of age-related neurodegeneration, such as Alzheimer's or Parkinson's disease, but also in schizophrenia, mood disorders, stroke, spinal cord injury, and drug addiction. This 3-year, $300,000 award began in 2001, with Larry Squire as committee chair. It would succeed the McKnight Investigator Award, which ended after its final funding round of 2000–2001. The initial announcement for this new award drew 272 proposals.

The McKnight Scholar Award, with its philosophy of funding young, talented investigators, was as relevant and as close to The McKnight Foundation's original goals in 1999 as it was at the beginning. The difficulty young scientists faced in obtaining federal funds had not eased, and no other openly available private philanthropy program focused on neuroscience, though some newer funds had small areas of overlap. Therefore, the directors in 1999 decided to keep the Scholar Award. They did, however, include it in their overarching attempt to reflect recent advances in basic neuroscience and foster translational work across the McKnight program. To this end, they rephrased the award announcement to attract to its roster a small number of investigators whose work bridges the gap between basic science and neurological diseases. The award's essence stayed unchanged. Importantly, application to all three awards is open—nominations are not needed.

Finally, the planning group seized on the conference as an opportunity to tie the manifold program changes together into a whole. To be held every year from 2000 on, the conference would be given greater emphasis as a tool to bridge the gap between basic and clinical neuroscience. Reflecting the two new awards, the conference now features a workshop on technology innovation in neuroscience and a workshop on a brain disorder selected annually by the EFN board. The latter serves to bring young basic McKnight scholars together with leading clinicians and experienced scientists on that particular disorder. Besides presentations, these half-day workshops schedule ample time for discussion and identification of the latest research opportunities. The goal is to entice young basic neuroscientists into identifying and tackling disease-related problems, perhaps even to strike up collaborations with other McKnight scientists.

This is a unique conference format in that it aims specifically to draw basic neuroscientists into studying mechanisms of disease. In this way, the conference has become an important vehicle for realizing the translational and interdisciplinary aspirations of the overall program. It also has fostered interactions and a growing sense of community among McKnight awardees.

In 2000, Corey Goodman became president of the EFN, Carla Shatz became vice president, and the board of directors turned over rapidly, as Larry Squire, Eric Nestler, Huda Zoghbi, David Julius, and David Tank joined one after another, with Stryer and Jessell rounding out the board.[2] In 2005, Shatz assumed the presidency, and Goodman became vice president.

These changes at the endowment fund coincided with changes at The McKnight Foundation. Michael O'Keefe moved on in early 1999, to be replaced by Rip Rapson, who was given the title of president and served until 2005. In 1999, Cynthia Boynton passed the board leadership on to her daughter, Noa Staryk, who assumed the title of chair. Staryk, a great-granddaughter of William McKnight, was the first of a new generation of Maude and William McKnight's descendants and their spouses to lead the Foundation's board. She remained as chair until September 2004, when her cousin Erika Binger took over.

The endowment fund was now positioned to weather these changes well, and the past six years have been a period of stability between the board's scientific representatives and The McKnight Foundation. They were marked by continued experimentation with the conference format and by the board's efforts to implement the new vision laid out in 1999.

---

[2] The EFN's tradition of stellar leadership continues. Goodman is president of Renovis, a neuroscience biotechnology company, and professor at the University of California, Berkeley. Shatz chairs the department of neurobiology at Harvard Medical School and has made key contributions to understanding the mechanisms that guide the formation of precise connections in the brain of higher mammals. Thomas Jessell is a professor at Columbia University College of Physicians and Surgeons and a world leader in the area of mammalian developmental neurobiology. Lubert Stryer, a professor at Stanford University School of Medicine, has invented a variety of fluorescent techniques for studying molecules and cells, and more recently co-invented one of the most important new tools in biomedical research: the gene chip. Larry Squire, recognized internationally for his research on the neurological foundations of memory, is a professor at the University of California School of Medicine in San Diego. Eric Nestler is a leading expert on drug addiction working at the University of Texas Southwestern Medical School. Huda Zoghbi, a professor at Baylor College of Medicine, applies the tools of modern genetics to understand brain development, neurodegeneration, and mental retardation. David Julius, a professor at the University of California at San Francisco, discovered the nerve cell molecules that determine how the body senses hot and cold temperatures. And David Tank, a professor at Princeton, has measured calcium ions in neurons and helped develop functional MRI imaging.

The two new awards took time to come into their own. Their stewards fine-tuned the award announcements to get across more clearly to neuroscientists exactly what kind of cross-pollination they were seeking to foster. Three years into each award, the directors felt that they were seeing the type and quality of proposal they were looking for broadly represented in the applications. In 2003, the McKnight Technological Innovations in Neuroscience Award had reached its 5-year point. Its committee had tracked the progress of the awards and felt that 50 percent had achieved their stated goals—a high rate given the riskiness of the projects. In 2005, the Neuroscience of Brain Disorders Award received 203 letters of intent from a wide range of basic scientists and clinicians interested in disease.

Increasingly during this time, the EFN faced situations where its awardees won simultaneous awards from other funders, such as the Howard Hughes Medical Institute, the John Merck Fund, or the Pew Charitable Trusts. The EFN considers each case carefully but is also guided by the desire to spread funds among as many deserving scientists as possible. In at least one case, a scientist chosen as a McKnight Scholar was allowed to defer the award until his other funding—which did not permit simultaneous awards—ended. If a McKnight Scholar receives the generous Howard Hughes award, however, the EFN asks the awardee to forego the remaining McKnight payments. Money already paid out need not be returned, and all awardees are invited to keep their status as McKnight fellows and return for the annual conference. While numerous other organizations now offer small grants for disease-oriented projects, the McKnight Scholar and Technological Innovations in Neuroscience awards remain among the very few that focus on basic neuroscience.

MCKNIGHT'S GIFT

The McKnight Endowment Fund for Neuroscience has evolved into the most prestigious private funder for neuroscience relevant to the brain. By any measure, it has come to be the crown jewel of William McKnight's legacy for brain science. But it was far from the only philanthropic initiative he undertook for brain research. In all, at least four organizations, not to mention their affiliates and grantees, are benefiting from his generosity as they explore the brain and its diseases, including disorders of memory.

- The McKnight Foundation, followed by its McKnight Endowment Fund for Neuroscience, to date has pledged nearly $90 million since 1977, mostly through awards to individual scientists.
- William McKnight had considered setting up a research program with the Mayo Clinic. In deference to his plans, The McKnight Foundation in 1978 approved $500,000 to endow the William L. McKnight–3M Chair in

Neuroscience; the 3M Foundation gave the same amount. Neurologist Andrew Engel, M.D., currently occupies the position.

- At the Medical University of South Carolina, William McKnight donated $1 million in honor of Boyle, which resulted several years later in the establishment of a Center on Aging and two endowed professorships.
- William McKnight's second wife, Evelyn, who died in 1999, came to share his interest in memory and used much of the fortune her husband left her to support memory-related research. She established the Evelyn F. McKnight Brain Research Foundation, now based in Orlando, Florida. It supports the more clinically focused research on aging and memory that The McKnight Foundation had decided to forego in favor of basic research. The primary activity of her foundation has been to establish the Evelyn F. McKnight Center for Age-Related Memory Loss at the University of Miami School of Medicine; the Evelyn F. and William L. McKnight Brain Institute at the University of Florida; and the Evelyn F. McKnight Brain Institute at the University of Alabama, Birmingham, School of Medicine. In 2006, a McKnight Scholar, J. David Sweatt (1990), assumed the Evelyn F. McKnight Endowed Chair for Learning and Memory in Aging at the University of Alabama School of Medicine.

In its emphasis on individual initiative and its principle that innovation depends on allowing people the freedom to fail, The McKnight Endowment Fund for Neuroscience stands as a tribute to William L. McKnight's spirit and a legacy of his practices at 3M. He could not have foreseen the organization his philanthropy made possible, nor could he have imagined how well it has served his goals.

AN INTERVIEW WITH JULIUS AXELROD

Julius Axelrod (1912–2004) followed an unlikely path to a Nobel Prize–winning career. He got his start in science by working as a laboratory technician at New York University. At $25 a month, it was the better of two entry-level positions open to Axelrod after he graduated from City College in 1933, in the depth of the Great Depression. (The other one would have been as a postal clerk.) Axelrod had hoped to be a physician, but medical schools had quotas for Jewish students at the time, and he was rejected. Next came an 11-year stint at New York City's Health Department developing assays to measure vitamins in foods. Axelrod first tasted actual research in 1946, when he joined Bernard Brodie at New York's Goldwater Memorial Hospital to study what caused the side effects of then-popular pain medicines such as acetanilide. He discovered that the body breaks down acetanilide into a toxic product, anilin, which he detected in his own urine, and a second one that is the actual analgesic. His first paper described this second compound, N-acetyl-p-aminophenol, a word that contains both the generic and a trademark name of the drug it led to (i.e., acetaminophen and Tylenol). From this first foray into pharmacology, Axelrod went on to pioneer the field of neuropsychopharmacology.

But first, in 1948, he joined the National Heart Institute, one of the precursor institutions of the National Institutes of Health (NIH). Over the next 10 years he investigated the effect and metabolism of stimulants such as caffeine, amphetamines, and ephedrine. Axelrod could not have known that his basic research would lay the foundation for the later development of synthetic adrenaline drugs. But precisely such a drug, Levophed, helped save his life 40 years later when he suffered a heart attack and his sagging blood pressure needed a boost to make him strong enough for bypass surgery.

In 1954, Axelrod was the first to describe a liver enzyme that would later become known as cytochrome P450, perhaps the best-known family of enzymes involved in metabolizing many drugs and dietary substances. But at age 42, with 25 papers to his name, many as sole author, he was turned down for a promotion because he did not have a Ph.D.

He took a year off, got his doctorate from George Washington University, and in 1955 left the National Heart Institute for the National Institute of Mental Health (NIMH). Thirteen years before the Society of Neuroscience would be founded, the NIMH was all of 6 years old, and Seymour Kety was directing intramural research at NIMH with the aim of establishing basic neuroscience as the biological underpinning of psychiatry. He gave Axelrod a small lab and a technician and told him to pursue anything he liked. Fifteen years later, Axelrod won a Nobel Prize. Axelrod's insight into reuptake of neurotransmitters as a way to control their activity was one of three basic advances that together earned the 1970 Nobel Prize in physiology or medicine; the other two were Ulf van Euler's discovery of the role of norepinephrine and Sir Bernard Katz's discovery of the nature of synaptic transmission.

The approach of valuing basic science over targeted research and then letting talent run free had been critical to Axelrod's best work, and it epitomizes the philosophy The McKnight Foundation would later develop. Axelrod's discovery paved the way for the development of modern psychiatric drugs, including anti-depressants such as fluoxetine (Prozac) and stimulants such as methylphenidate (Ritalin). He then studied endorphins and melatonin, a hormone involved in cir-cadian rhythms, among other topics.

Despite his late start, Axelrod spent 50 years as a research scientist, retiring from his lab at age 80. He resisted offers to head departments and otherwise advance his career. He preferred to stay at the bench without the pressures of writing grant pro-posals and the responsibilities of administration. "My greatest satisfaction is that my basic work led to treatments for pain and depression," he said.

Through his former postdoctoral student Leslie Iversen, an eminent pharma-cologist at University of Oxford in England, Axelrod "fathered" two McKnight awardees: Thomas M. Jessell (Investigator Award, 1985; current board member of The McKnight Endowment Fund for Neuroscience) was a postdoc with Iversen, and Marc Tessier-Lavigne (Scholar Award, 1991; Investigator Award, 1994) was a postdoc with Jessell. "This makes Marc my great-grandpostdoc," Axelrod quipped.

Axelrod sat down with Gabrielle Strobel on September 29, 2004, at his apart-ment in Bethesda, Maryland, where at age 92 he lived independently. Asked what made life rewarding, he offered this unsurprising advice: "Stay at the bench." Axelrod died on December 29, 2004.

Strobel: **McKnight Awards are among the most prestigious in neuroscience. Many of today's distinguished neuroscientists have won one early on in their careers. The awardee list reads like a who's who in neuroscience, with 47 members of the National Academy of Sciences and five Nobel laureates. What made this success possible?**

Axelrod: Above all, the choice of people. We had high-quality people on the scientific advisory board and the selection committees. They had done lots of reviewing before and were able to pick the best applicants. [The people selected] became outstanding in their field. Our guiding principle was that we would not target funds to Alzheimer's [or any other] disease. This broadened the field. We wanted basic information about how the nervous system retained information. In neuroscience, that was radical at the time, but we thought it was doable. Above all, we sought imagination and great ability in the awardees; we cared less about specific details of their proposals. Some awardees had track records; some did not yet but made their marks later.

In setting these guidelines, who helped you the most?
I got a lot out of talking to Seymour Kety, who led the NIMH intramural program at the time, and Edward Evarts, the head of my lab. They both took an intense interest, because this program allowed them to avoid the restrictions imposed by reviewing [for federal grants]. Russ Ewald had a great influence. He was an ex-minister, not a scientist, but he had keen judgment about people. I could tell by the way he conducted our meetings. What pleased him the most was that we really got the most notable people in the field.

For 5 years prior to your involvement, Mr. McKnight had supported research led by Dr. Edwin Boyle at the Miami Heart Institute, called the McKnight Memory Loss Program. What do you remember about that?
Edwin Boyle was not the expert he pretended to be. McKnight had memory loss that was probably Alzheimer's, although people did not call it that until some time after Robert Katzman published about dementia using this term, in the late 1970s.

The work involved hyperbaric oxygen. Older people with memory problems were placed in pressure tanks.
Mediocre science at best.

How could Boyle conduct this work for 5 years without oversight?
Because he was convincing. He was a postdoc in Chris Anfinson's lab at NIH when Chris was already a Nobel laureate, and that was his claim as a scientist.

A philanthropist funding a scientist who is also a friend — was, or is, that a fairly common arrangement?
Yes, but it is not a wise thing. Emotions take over. You need peer review. You need authority, people with integrity.

How did Mr. McKnight react to the outside review of Boyle's research?
I don't know if it was Mr. McKnight. It might have been his daughter, Ginnie [Virginia McKnight Binger], who realized he was wasting his money. She then came to our meetings. She was very much interested in how we conducted our business. I never met Mr. McKnight.

How do you remember the first years?
One day, I got a call from Russ Ewald saying he'd like to set up a funding program for memory research. That was after they had realized Boyle was the wrong person. Ewald asked my advice. I said that you need a scientific advisory board [SAB]. I suggested three types of award: one for young assistant professors who just got their independence, one for investigators who are up-and-coming, and one for people who have already contributed a great deal. They phased out this

Senior Award in 1999, and in hindsight I really prefer giving money to young people. Evidently Ewald must have agreed with that. He asked me to become chairman of an SAB. We chose people who are very much neuroscientists, who had particular expertise in the neuroscience of memory. We did not believe in psychological memory research but instead tried to push the field toward molecular science.

I assembled the SAB with Kety's help. He was such a great scientist. He was the first to measure blood flow in the brain, which became the basis for PET scanning. He also was the schizophrenia czar at the time. Looking through a registry from Denmark, where they record the lives of people for hundreds of years, his group established that schizophrenia was at least 50 percent genetic and that it was polygenic. While leading the intramural program at NIMH for many years, he established biological psychiatry in a field that was heavily dominated by Freudians.

How did you select awardees?
We decided against the NIH way of making proposals, where you have to write long proposals of 30-plus pages. We asked for only three pages. I recommended that we solicit grant proposals individually, rather than advertise in *Science* and *Nature*. That would put the onus of finding the right candidates on us, but otherwise we'd be deluged and need a whole staff. Then we picked awardees personally, and the people on the SAB knew them. We gave proposals to particular experts depending on their nature, to have a first review and a second review. We met several times a year to discuss the proposals. That gave us a highly select group. We probably missed some good people, but all the ones we did fund were good. What foundation can say as much?

The awardees all got NIH grants, but we gave them extra. Creative people need that because NIH money is restricted. We gave generous money carte blanche. NIH requires preliminary evidence for your proposal, and this discourages novel, wild ideas. We encouraged [such ideas]. The people we chose became distinguished in their own right in all areas of neuroscience; we just helped them get established. So I think The McKnight Foundation took the right course. It used a limited amount of money to great effect.

How unusual was McKnight funding program at the time?
NIH was the main source of money. Besides that, I remember only the Sloan Foundation, which operated in a similar way but did not focus on neuroscience. The mutual fund manager Jack Dreyfus supported research, but that was targeted to his own ideas of what he thought caused depression. I can't think of any other funders like McKnight.

Whom did foundations fund, mostly?
Existing institutions.

What would have been a typical funding strategy for a foundation interested in neuroscience at the time?
Give some to an institution, some to the neurology chairman, some to individuals. I don't know much about foundations.

In the early 1970s, was there even much biological research yet on the basic science that underlies memory disorders?
Yes, some of that existed. One of the founders was Eric Kandel, who worked with sea slugs. He established various molecular routes leading to the formation of simple memories and manipulated them. Another authority was James McGaugh. Both were funded by McKnight early on.

Did the McKnight program play a significant role in creating this science?
It allowed it to flourish. The SAB people we had gave us stature, and they were molecular. We'd invited no psychoanalysts on it. The reason Kandel won the Nobel Prize was this molecular biology of memory. The foundation also suggested we hold annual meetings, where the recipients gave talks. This created a sense of community for the fellows but also a forum for this nascent field.

Why did Ewald call on you? Because you were the reigning Nobel laureate on neurotransmitters?
Probably. I do not know who recommended me, maybe Fred Plum.

What was already decided by the time you became involved? What existed, and what did you newly conceptualize?
They had decided to end Boyle's program and to begin a neuroscience program. I wrote a proposal and gave it to Russ. It proposed funding people who had wild ideas that you don't know are going to work. Selecting three types of people, the beginners, the up-and-comers, and the established, that was my idea. Fred wanted to set up a center of excellence at Cornell. That was up in the air when I came on. I thought it was a bad idea.

Why?
I thought the funding should be broader, because the field's most important assets are experts throughout the country. Money should go directly to them, not to administrators at an institution. Russ had a good instinct about funding. He had done it for orchestras and other institutions. He sensed the funds should not be restricted to an organization and was receptive to my argument. My basic idea was to find people who are very promising and then allow them to do whatever they want. They have to write a proposal, but once they have the money they could do as they pleased, even gamble it away in Las Vegas. One person did actually abuse it privately.

Besides the institutional aspect, a center of excellence would have meant targeted research. I thought that, too, was a bad idea. Kety put it as a slogan:

"Excellence, Novelty, Not Conspicuous Relevance." If it's too obviously relevant to disease, it's probably mediocre science. In my letter to Russ I recommended that, rather than support an institution, they set up a peer review like NIH and let the peers decide whom to fund. It was my decision in the end, because Russ asked me to assemble the scientific committee that would make this call.

Other organizations that fund research into brain diseases have not always had the impact they could have had because they focused on medical studies early, when the science was not ready for that. Even today, many foundations concentrate squarely on one disease. Given that Mr. McKnight was driven by his concern about memory loss, his foundation easily could have decided to fund medical research targeted specifically to this condition. Why was it so important to you to promote basic science and excellence instead?
I was inspired by Julius Comroe and Robert Dripps's classic report, "The Top Ten Clinical Advances in Cardiovascular-Pulmonary Medicine and Surgery 1945–1975."[3] It established how important basic science is to medical progress. It's still relevant today. Comroe and Dripps had asked clinicians and physiologists to list the most seminal contributions that led to diagnosis, treatment, and prevention of heart disease. Nearly 80 percent were not originally directed at heart disease. (Incidentally, this group contains one of my papers.) Only 15 percent were targeted. I wish someone did this kind of study for neuroscience. If you make it too targeted, you waste a lot of money. I am sure that is true today. Our decision not to fund targeted science created dozens of scientists who contributed immeasurably to neuroscience, even if perhaps they did not work directly on Alzheimer's.

Did it take nothing more than your letter to persuade Ewald? No impassioned arguments at meetings, heated phone discussions? No drama at all?
No. Russ was a foundation man with good judgment. He was persuaded quickly. He made his decision, then called me and asked me to round up the SAB. Next we formed a very distinguished ad hoc committee. We all agreed on this, and Russ and Ginnie and Jim Binger [Ginnie's husband and a director of the foundation] took our advice at our first meeting. The brochure accompanying their first announcement spelled it out clearly.

What won over Mr. McKnight?
He was already sick; that is why the Bingers wanted this program. They trusted Ewald and had given him full authority, so it was his decision to go with my advice. By the way, he was full of Midwestern humor, but that did not mask the fact that he had great power.

---

[3] Department of Health, Education, and Welfare, 1977; first published in *Science* as "Scientific basis for the support of biomedical science," 1976; 192:105–111.

Did anybody resist your priorities of basic science/young scholars/hand-picked excellence? No disagreements about direction?
No. I picked people who knew the value of basic science. Some favored more targeted research, but we decided to wait with that.

Did your own professional experience play into setting these standards? Had you benefited from someone "discovering" a bright young guy who deserved a chance? Or, on the contrary, had you felt impeded for lack of unrestricted funding?
My background at the NIH shaped my thinking. Its founders were distinguished, and they set up NIH in a way that made it a prestigious organization.

Before the NIH, the amount of money Congress gave to research was minuscule. Actually, John D. Rockefeller was the leading funder of biomedical research, but he did it mostly through Rockefeller University. Setting up peer review was central to NIH's success, so I took this part of NIH as my model. I also had complete freedom; that is why I stayed at NIH.[4] Kety told me to pick any topic I wanted to work on, as long as it's novel and important. This freedom, too, I adopted for the McKnight program. What I did not take is the bureaucracy, and the "make sure it's going to work" approach. They [at NIH] are very conservative; that is one of their faults. For McKnight, we adopted the "take a chance on a bright person" approach.

You never feathered your own nest; that's worth a mention for someone who worked for many years so close to arguably easy money.
They never funded my research; that would have been a conflict of interest. They knew I like art so they gave me a Matisse etching as a gift when I left.

Today, most established biomedical scientists interact with philanthropic foundations in some way. That was not yet so in the 1970s. What motivated you to spend time on such an unusual activity?
I felt it was my duty not only to do research but also to spread good research. One way you can do that is to ask other respected scientists to find and enable bright young people. I felt pride in doing this. It is the recognition of your stature that you are asked to do those things. I am not a "macher." I never became an officer of the Society for Neuroscience; I was not interested in power. At NIH, I was never even a lab chief. Section chief was my biggest position, where I had four postdocs at most. I did not want administration. I hate to make unpleasant decisions or to look for money, and in that sense the intramural program was right for me. I don't think ahead for years to plan out huge grants. Usually I did not know what I would do next week. I followed my nose. I wanted to pursue

---

[4] According to the NIH website, the NIH budget expanded from $8 million in 1947 to more than $1 billion in 1966. Between 1955 and 1968, NIH director James Shannon presided over the spectacular growth that is now remembered as "the golden years" of NIH expansion.

one idea after another, as they came to me. To have success in science, you have to have a lot of ideas, because most of them fail. So the McKnight program was a way of giving back that suited my personality.

You do not come from a privileged family that destined you to an Ivy League career. You grew up in tenement housing in New York City as the son of a Jewish basket maker. What in your background influenced your ethos about science?
In my Jewish family, medicine was considered a princely pursuit. I really wanted that but did not get into medical school. Second, reading *Arrowsmith* by Sinclair Lewis and *Microbe Hunters* by Paul de Kruif left a deep impression on me. Reading about the lives of the scientists portrayed in these books, I thought, "That is the kind of life I want to lead." I went to City College, which was free at the time, but most students were bright and highly motivated. City College generated eight Nobel laureates and many distinguished biochemists, all from the background I had. Being a scholar is highly prestigious for Jewish people. Among Shtetl Jews, marrying the rabbi's son was the best a girl could do. And think back to the *Fiddler on the Roof*: no matter how poor, Tevje was learned.

Of all your research, which discoveries mean the most to you?
Clearly, my most obviously important work is the neurotransmitter uptake work, for which I won the Nobel Prize. But I care as much about a finding I got little credit for. I described a class of enzymes in the microsome compartment of liver cells that metabolizes drugs and foreign compounds. Now it is called P450 and one of the most studied of all enzymes. But I described it first. Most dear to my heart is my work on the pineal gland and its hormone melatonin, because all the experiments I did on it actually worked. It was my personal antidepressant. Whenever I got frustrated about other things not working, I turned to the pineal gland. Every success in science is a great uplifter, because most experiments actually fail.

Did you ever have disagreements with the McKnight people? How do you look back?
Disagreements, no. They are honest, straight people. I enjoyed my work with them. I feel pleased that the program turned out the way I hoped, that it would distribute the money in the best possible way to support broadly basic research. To be responsible for developing so many distinguished future neuroscientists gives me a great deal of satisfaction.

Chapter 2

A Budding Field Begins to Flourish

HOW THE McKNIGHT FOUNDATION HAS NURTURED NEUROSCIENCE

In 1976, when The McKnight Foundation's directors decided to develop a neuroscience research program, they were putting their faith in a field that was still young. Neuroscience was just beginning to emerge from the folds of academic departments of anatomy, physiology, or psychology and to assert itself as a discipline in its own right. The Society for Neuroscience had been founded 6 years earlier with 500 members. By 2006, it had more than 37,500 members, living proof of the field's explosion into manifold subareas. As it has grown, it has also matured.

In the 1970s, however, neuroscience was still being conducted narrowly. Biochemistry had been making inroads into the exploration of the brain for some time, but modern genetics and molecular biology had not yet penetrated brain research, as they would in a major way only a few years later.

The leaders in neuroscience were neurophysiologists, neuroanatomists, and physiological psychologists. They did imaginative science but were limited in the tools and concepts available. Many performed their experimental work on invertebrates, including snails and leeches, crabs and lobsters, grasshoppers and tobacco hornworms, even cockroaches. These were favorite preparations because they made it possible to obtain large, identifiable neurons that could be poked with recording electrodes and would respond in ways that could be compared from one experiment to the next. This was also why the electric eel found itself in neuroscience laboratories in those days. Vertebrate preparations for the most part tended to be the less complex, accessible ones of the peripheral nervous

Research Funding in Neuroscience

system, such as the synapses with which motor nerves end onto muscle fibers, ganglions of the autonomic nervous system controlling the heart, or the dorsal root ganglion at the boundary between the peripheral nervous system and the spinal cord. In addition, scientists interested in behavior studied rats and monkeys with techniques such as lesions, single-cell recordings, and drugs—all with the aim of understanding the neural substrates of behavior.

Typically, neuroscientists used electron microscopy to visualize fine details of neural tissue that was chemically fixed. Some were delineating circuits in the brain with dyes or radioactive tracers. Others were identifying neurotransmitters, culturing neurons, and dripping neurotransmitter solutions on them. Some were sticking glass electrodes into neurons and recording currents from them, while others were studying the effect of drugs on the electrophysiology of neurons and the behavior of experimental animals.

To be sure, there was plenty of innovation across neuroscience at that time. To cite just two examples, by the mid-1970s Eric Kandel had already coaxed from the sea slug *Aplysia californica* a lucid description of fundamental mechanisms of learning, such as sensitization and habituation to a stimulus.[1] In 2000, this work would result in a Nobel Prize. Even then, he and other leaders realized that functional and structural changes in synapses underlie those behaviors, but they had no means of addressing precisely what the changes were. David Hubel, later a McKnight Senior Investigator, and Torsten Wiesel, later president of the Endowment Fund for Neuroscience (EFN), were breaking new ground on vision with their electrophysiology studies in the visual cortex of cats and monkeys, research that in 1981 would be recognized with a Nobel Prize. But even this stellar work remained physiological at the time.

A more profound revolution in the field occurred after molecular genetics and recombinant technologies had entered the scene and were beckoning to be applied to neuroscience problems.

Broadly speaking, researchers knew the basic phenomena of the brain and nervous system—the axon, neuronal structure, and synaptic transmission. Many of the big questions in neuroscience had been framed and appreciated—that is, how do axons find their way, how do they make synapses to wire up a brain, how does neurotransmission truly work, and how does it generate behavior? The stumbling block was that the field had difficulty approaching them in a mechanistic way. Without that, it could not rigorously approach questions about mechanisms gone awry in disease, either. "The piece that was missing was the identity of the molecules you were working with. Because we did not know what these molecules were, the level of neuroscience research remained descriptive," recalled

---

[1] Kandel helped shape the EFN, and in 1977 and 1980 he received Senior Investigator Awards. He was a board member for 12 years.

Sam Barondes, a founding board member of the McKnight neuroscience program and longtime EFN president.

Biochemistry was already helping neuroscientists, in that it enabled them to turn pharmacologic phenomena into concrete biochemical entities — proteins that one could isolate and manipulate. For example, Arthur Karlin had discovered in 1966 that characteristic current changes across a neural membrane in response to nicotine were the result of a particular type of acetylcholine receptor protein. (Incidentally, 20 years later, Karlin received one of a handful of discretionary awards the McKnight Endowment Fund has granted, to purchase equipment for the Summer Course in Neurobiology at the Marine Biological Laboratory in Woods Hole, Massachusetts. This legendary summer camp for neuroscientists at the time featured as faculty German Nobel laureate Bert Sakmann and a young Roderick MacKinnon, who would soon do his own Nobel Prize–winning work.)

A more profound revolution in the field occurred after molecular genetics and recombinant technologies had entered the scene and were beckoning to be applied to neuroscience problems. Over the next two decades, they would transform the field completely.

Neuroscientists were divided about this change. For his part, Kandel seized the new opportunity and expanded his work to include molecular biology. Other former physiologists also became leaders in molecular neuroscience. Yet many of their colleagues were deeply skeptical. "Some people said the molecular level would be impossible to decipher. Some said that molecules were not important, that only the phenomenon itself was important," said Gerald Fischbach, currently dean of the College of Physicians & Surgeons at Columbia University, who was involved with the EFN for 12 years. Others felt that a focus on molecules would detract from the importance of integrative and systems neuroscience.

In hindsight, no one today seriously disputes that molecular approaches to neuroscience problems were key to being able to manipulate the processes and in this way understand the phenomena better. Molecular approaches transformed neuroscience from an observational science to a truly experimental one. To cite but one example, Carla Shatz and Michael Stryker had trained as neurophysiologists in the visual system, but then incorporated molecular explanations into their study of how the visual cortex forms and changes with experience. That it changes was part of Hubel's and Wiesel's original discoveries, but figuring out the mechanisms of how it does so depended on molecular approaches.

The McKnight program had its start right at the dawn of an extended period of growth and continuous change in the field.

But self-evident as the accomplishments of molecular biology are in hindsight, in the late 1970s and early 1980s, pushing for its inclusion into neuroscience took vision and a taste for risk. This is where the McKnight neuroscience program made its mark. The McKnight program had its start right at the dawn of an extended period of growth and continuous change in the field. The program's funding record shows a legacy of picking the best neuroscientists at the time in their careers when they were young and hungry to chart a new course in science, or when they were more established but about to embark on a major change in direction. It then gave them unrestricted funds that allowed many of them to do their best work and to become the leaders who helped realize the coming transformations.

LAB AND CLINIC, WORLDS APART

Before reviewing some of these research advances, it is useful to consider the situation with regard to translational science in the early years. After all, William McKnight's goal was not to advance science for its own sake but to foster a better understanding of diseases that destroy memory and cognition.

In the late 1970s, The McKnight Foundation decided to take a long view on this goal. Executive vice president Russ Ewald and the directors concurred with their scientific advisers that it would be wise to forgo funding targeted research for the time being and to invest instead in fundamental exploration, until neuroscience had a better grasp of brain function. At the time, the foundation could not have foreseen the future rise of the biotechnology industry, nor the manifold close connections this industry would forge with academic neuroscientists beginning in the 1990s. These connections are a direct result of the growing body of basic research McKnight helped foster.

Neither did the foundation know at the time that its program, focused as it was on basic science, would nonetheless produce some of today's leading innovators in biotech and pharmaceutical companies, such as Corey Goodman at Renovis, Richard Scheller and Marc Tessier-Lavigne at Genentech, David Bredt at Eli Lilly and Co., or Mark Gurney at deCODE genetics—all McKnight awardees. What they saw was that precious little substantive exchange was occurring between basic neuroscience on the one hand and neurology and psychiatry research on the other.

Why the silence? Today, the relevance of brain science to brain diseases seems plainly obvious. But in the 1970s, neurologists generally did not think of fundamental neuroscience as informing "their" diseases in important ways. Many clinicians knew little about what the basic researchers were doing. Moreover, few effective drugs were available to neurologists. It is worth noting, however, that the introduction of L-dopa for the treatment of Parkinson's disease in the late 1960s was a spectacular exception that was beginning to raise the image of neuroscience in the minds of neurologists and change their expectations of it.

For their part, basic neuroscientists believed their work would ultimately be relevant to neurologic disease, but the connection did not begin to be made in a strong way until the 1990s, when translational science expanded. Before that, investigators' grant applications to the National Institutes of Health tended to end with short sections in which they stretched their imaginations to make a case for the human relevance of their proposed work. Meant to increase the application's prospects, and required by the proposal guidelines, these statements often did little more than pay lip service to the goal of human application. Now the body of evidence to which these studies contributed is indeed relevant, but the individual studies at the time generally were not.

The history of the biotechnology industry reflects this separation as well. In the early 1990s, several small companies made early forays into treating neurologic conditions with agents derived from the initial results of cellular and molecular studies, but they failed. There simply was insufficient basic knowledge on which companies could draw in designing therapies and trials. Prime examples include human tests of growth factors and cytokines to treat neurodegenerative diseases. Despite early setbacks in the clinic, these proteins are still being studied. They may yet yield therapies in the future as their important interactions with other proteins become better known and drug delivery techniques improve.

There was also a sociological barrier to close interaction. Basic scientists tended to view with disdain those colleagues who tried to work on a neurologic problem. They knew that it was difficult to do rigorous basic science on a clinical problem, especially in the absence of a good technical handle. "We did not have much dialog with the clinicians. Too little was known and we did not have a shared language," said Corey Goodman, president of The McKnight Endowment Fund from 2000 to 2005 and a longtime member of the board who himself left an academic position at the University of California, Berkeley, to head biotechnology company Renovis.

Focusing on basic investigation would build up an account from which later translational efforts could draw.

That lack of common ground between bench and bedside is why Julius Axelrod, who in 1970 had won a Nobel Prize for his exploration of synaptic transmitters, advised The McKnight Foundation to focus its initial program squarely on basic investigation. This would build up an account from which later translational efforts could draw. It is also why the foundation rejected as well intentioned but premature a funding proposal for an integrated basic-clinical research center.

In psychiatry, the situation was even worse than in neurology. Psychoanalysis strongly influenced clinical practice. Some psychiatrists had accepted the advent of antipsychotic drugs with reluctance and suspicion. The leadership at the

National Institute of Mental Health (NIMH) and at some institutions recognized the promise of molecular biology for psychiatric research, but many academic psychiatrists had less sympathy for a biological approach to diseases of thinking and mood. The idea of doing molecular experiments implied to them that the talk therapies they were practicing might be of limited use. It also implied that people's behavior was at least in part genetically determined.

Through the 1960s, the idea that genes distinguished people as much as learned behavior was still resisted. A baby was considered a blank slate, and the influence of the environment and caretakers was thought to shape a child's personality. In child rearing, the mantra was that boys play with trucks and girls with dolls because mothers encourage gender stereotypes, not because of inherent preferences. In medicine, psychiatrists believed that a person's upbringing determined his or her adjustment in society. By that logic, aberrant thinking as seen in schizophrenia was attributed to poor parenting. Psychiatric conditions were especially likely to be interpreted in terms of prevailing social thought, perhaps because they so often affect adolescents and young adults, whereas age-related neurodegenerative diseases erode an established personality. Little rigorous biology was being done on psychiatric illnesses at the time, leaving a void for social determinism to fill.

In both neurology and psychiatry, the McKnight program engaged advocates of molecular genetic investigation right from the start and called on the best minds available. One of its founding scientific advisers was Seymour Kety, the first leader of the NIMH and a mentor of Axelrod's. Kety had fought to ground psychiatric research in biology. He lifted a burden of guilt off parents' shoulders when he discovered in 1971 that schizophrenia has a strong genetic component.[2] Kety backed Axelrod's stance on basic and molecular research. So did Barondes, Kandel, and Solomon Snyder, all trained psychiatrists who advocated for a biological approach and for bringing neuroscience to the clinical arena; Torsten Wiesel was sympathetic to this view as well. At present, Eric Nestler continues the tradition of having a trained psychiatrist and neuroscientist on The McKnight Endowment Fund board.

Today, advances made in neurogenetics, molecular biology, and brain imaging have altered the view of clinicians—though perhaps less so of the public at large—about the importance of biologic study of diseases of the mind. Molecular exploration in both neurology and psychiatry has become dominant as researchers try to emulate the progress that has been made in some diseases, such as Alzheimer's (see case study on page 69).

Progress has been slower in psychiatry than in neurology, but thanks in part to The McKnight Endowment Fund, techniques, talent, and better animal models

---

[2] This discovery, along with pioneering work on a forerunner of modern brain imaging, earned Kety the Albert Lasker Award for Lifetime Achievement in 1999, shortly before his death the next spring.

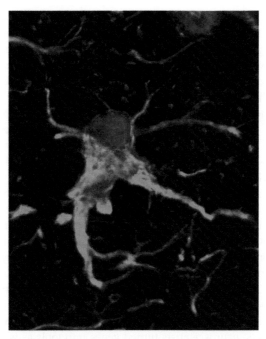

Paul H. Patterson received a 2003 Brain Disorders Award to create a mouse model of schizophrenia. This image shows how neuronal injury rouses glial support cells called astrocytes in the adult mouse brain. Courtesy of Sylvian Bauer and Paul H. Patterson.

are in place there too. For example, Mark Gurney, one of the scientists who identified neuregulin as one of the few substantiated risk factor genes for schizophrenia, received a McKnight Investigator Award early in his career (1985).[3] Two-time awardee Lorna Role (Scholar, 1988; Investigator, 1994) today studies the protein's function, and McKnight Brain Disorders Awards to Pat Levitt (2002), Paul Patterson (2003), and Jill Morris (2006) are dedicated to understanding different aspects of this complex disease. "The impact of molecular science on psychiatric diseases is around the corner," predicted Kandel.

DECADES OF EXPANSION

Even a cursory look at the course of neuroscience since the inception of the McKnight program reveals that molecular biology has revolutionized the field. "At McKnight, we had our eyes on molecular science right from the start. Our

[3] Gurney is also known for making a widely used mouse model of ALS.

vision was clear and accurate," recalled Barondes. The effect has been to enlarge both neuroscience overall and molecular biology itself. In identifying genes, molecular biologists opened up large new areas of inquiry for themselves while revolutionizing what neuroanatomists, neurophysiologists, and behavioral neuroscientists can do. These latter disciplines are relevant today because scientists can conduct their research with more sophistication, making links between neural circuits and function in terms of the genes that drive them. This has become possible thanks to specific probes and other tools, which were invented as a consequence of scientists' identifying genes and learning how to manipulate them. There was also a confluence with advances in imaging technologies at the molecular, cellular, and whole-brain level.

Together, these developments have provided scientists with the genetic tools to introduce tracers into nerve cells so they can be made visible, even in the living animal. Others use probes to localize receptors and to target drugs to specific receptors. Scientists can light up fluorescent proteins in particular spots within a neuron, or color the flow of ions in response to neural activity. This enables researchers to monitor subcellular functions of neurons in culture and even to probe particular circuits in the intact brain when that circuit is actively involved in a particular behavior.

In short, by first using biochemistry to understand genetics and then fashioning useful reagents, neuroscientists turned genes into tools as well as objects of study. They can modify genes, put them into an organism like a mouse, take them out, and in this way push the study of brain function and dysfunction to a deeper level of analysis. The completion, more recently, of the human genome as well as the genome of a number of animals, has given researchers a trove of information. They draw from it in their efforts to identify how genes interact to control a brain function of interest. This helps people who are trying to uncover mechanisms of action that underlie clinical problems in neurology, psychiatry, even traumatic brain injury.

This cross-fertilization of techniques has had the added consequence of blurring the distinctions between the different methods of neuroscience. Investigators nowadays define themselves less by their training—be it in anatomy, physiology, biochemistry, molecular biology, or psychology—than by their investigative role as neuroscientists. They simply bring different methods to bear on their subject.

"Molecular biology, genetics, imaging, and cognitive neuroscience all have advanced impressively," said McKnight Endowment Fund director Tom Jessell, who currently chairs the Scholar Awards Committee. "The big questions in neuroscience we understood back then, but now we have a real opportunity to approach them."

To be sure, the process is far from complete. On the basic science side, fundamental gaps in our understanding of how neural systems work remain to be bridged. On the clinical side, the transfer of basic science to the bedside has only begun to be made in an effective way. While molecular genetic science has yielded diagnostic laboratory tests for Huntington's and rare inherited forms of other neurologic diseases, it has not done so for their more common sporadic forms or for psychiatric diseases. Likewise, molecular genetic science has not yet produced new mechanism-based drugs for these diseases. But the process toward these goals is well under way, as shown in the selected examples that follow.

A NOTE ABOUT THE EXAMPLES

Through 2005, The McKnight Foundation and The McKnight Endowment Fund have granted more than 350 awards. Assembled in a diagram, the lineages of who trained whom and who influenced whose course of research would create an image much like that of an evolutionary tree. This tree of lineages and cross-talk would pervade the development of neuroscience throughout its exponential growth. As scientists follow a pioneer and join in the exploration of a newly opened research area, this area expands quickly. It draws on many influences from other areas as ideas and findings converge and techniques are adapted.

Tracing the contributions of all McKnight scientists to the field of neuroscience is beyond the scope of this publication. Likewise, this essay does not chronicle the contribution of many other great scientists who have helped establish the biochemical and molecular analysis of neural phenomena but were not involved with McKnight—such as Jean-Pierre Changeux, Arvid Carlsson, and Floyd Bloom, to name but a few. Instead, this essay showcases a few selected examples of leaders who came through the McKnight program and seeded new fields of investigation. The fields they opened interconnect at multiple levels with others, and, consequently, the story lines overlap at times. For example, developmental principles inform the study of adult brain function, and ion channel physiology informs our understanding of a growing number of brain disorders, ranging from epilepsy to mental retardation.

SAMPLING THE SCIENCE

Consider first the topic of how developing axons recognize where to grow and how to wire up the brain. This is an area to which many McKnight awardees have contributed. The foundation began funding this question in its very first year of giving awards, 1977, with a Scholar Award to Urs Rutishauser. Trained as a biochemist, Rutishauser had the year before identified a protein called neural cell adhesion molecule, and he then plunged into neuroscience to establish the protein's role in sticking neurons together as they build a tissue. In 1978, Louis

Reichardt became a McKnight Scholar. Now head of the neuroscience program at the University of California, San Francisco, Reichardt was one of the first neurobiologists to seize on the new technologies of monoclonal antibodies and recombinant DNA. He has since become renowned for exploring how neurons recognize molecular clues on the surface of cells and transmit signals to their command-and-control center in the nucleus.

In 1980, the Scholar Award selection committee picked Corey Goodman, who would become a world leader in neuroscience. Back then, Goodman was a young faculty member at Stanford University, studying neural development in the grass-hopper, a seemingly odd topic that funders other than McKnight did not find appealing. "How some locust wires up its nerves might not have seemed imme-diately relevant to human brain diseases," Barondes recalled, "but he clearly was brilliant, and we were wise enough to realize that this work would be important in the future."

For the next few years, Goodman pioneered approaches for studying neural development in simple nervous systems. They were attractive because the scien-tist could name and identify individual neurons from embryo to embryo, follow the trajectory of their growth cones (i.e., the hand-shaped leading edge of an outgrowing axon), and trace where and how they make synapses.

In parallel, Lynn Landmesser studied the development of motor neurons in the limbs of chick embryos. At the time, there was much uncertainty about how specific axons grow toward their proper targets during development, and Goodman's and Landmesser's work showed that axons select particular path-ways and grow along them in a standardized and stereotypical pattern, not at random.

On a separate track, behavioral genetics of these simple systems developed at the same time. McKnight awardee Seymour Benzer (Senior Investigator, 1994; Brain Disorders, 2004) pioneered fruit fly genetics, and Sidney Brenner did the same with roundworms. They observed mutant animals with odd behaviors or learning deficits and developed methods to identify the mutated genes causing the defects. This and other work convinced Goodman that molecular biology could help explain how neurons established circuits, and in the mid-1980s, with a McKnight Investigator Award, he moved his research from the grasshopper to the genetically more tractable fly. In the following years, his lab defined a catalog of genes involved in brain wiring, showing that the specificity seen in insect neurons with prior, descriptive methods had a corresponding molecular basis. Goodman's lab introduced whole classes of proteins that subserved the function of guiding a growing axon along its proper path and controlling how it sets up synapses of the proper size and strength.

Around the same time, another preeminent leader in developmental biology, Thomas Jessell, received McKnight support. A 1985 McKnight Investigator, he has trained acclaimed leaders in the areas of pain perception and chemotropic proteins, such as David Julius (Scholar, 1990; Investigator, 1997, and current board member) and Marc Tessier-Lavigne. Tessier-Lavigne (Scholar, 1991; Investigator, 1994), a prominent neuroscientist, has advanced the understanding of molecular guidance in neural development and regeneration to new levels with studies of chemoattractant and repellent proteins that are universally found in vertebrates, including humans.

Jessell himself is perhaps best known for taking to the molecular level Landmesser's anatomical description of particular classes of motor neurons in the spinal cord. These classes of neurons send out axons that make specific choices in innervating particular parts of muscle. Jessell discovered and characterized the signaling proteins and transcription factors that specify the fate of naïve progenitor cells in the embryonic spinal cord as they mature into different classes of motor neurons and interneurons. He also defined the molecules that

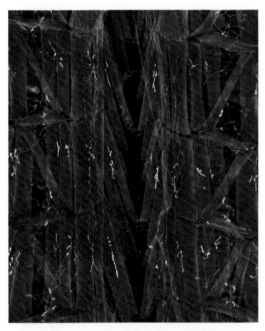

The *Drosophila* neuromuscular synapse, with muscles in green, nerves in blue and synapses in red, from the laboratory of Aaron DiAntonio (Scholar, 2002). DiAntonio studies the molecular sequence of how synapses grow so they can establish circuits and support learning. Reprinted from the cover of the *Journal of Neuroscience*, 24 (November 2004); in conjunction with an article by Daniels, Collins, Gelfand, Dant, Brooks, Krantz, and DiAntonio, "Increased Expression of the *Drosophila* Vesicular Glutamate Transporter Leads to Excess Glutamate Release and a Compensatory Decrease in Quantal Content," 10466–10474, © 2004 by the Society for Neuroscience; used with permission.

enable those neurons to make selective connections with target cells so a functional locomotor circuit can form.

In the past 5 years, scientists' understanding of axon pathfinding and connectivity has become quite sophisticated.

Goodman's and Jessell's work triggered a wave of research by others. Over a 25-year period, younger scientists who trained with them have made a steady stream of contributions to the growing area of neural specification and axon guidance. Indeed, in the past 5 years, scientists' understanding of axon pathfinding and connectivity has become quite sophisticated, at least for the most heavily

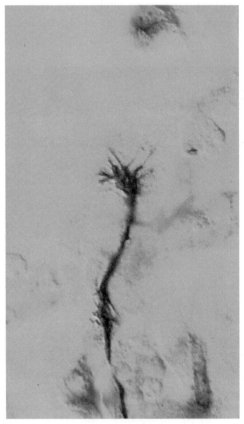

David Van Vactor (Scholar, 1997) studies the developing brain. Here, a fly neuron's growth cone (dark, fingered structure) explores the surface of a muscle to make connections that will enable coordinated movement. Reprinted from *Neuron*, 22; Wills, Marr, Zinn, Goodman, and Van Vactor, "Profilin and the Abl Tyrosine Kinase Are Required for Motor Axon Outgrowth in the *Drosophila* Embryo," 291–299, © 1999, with permission from Elsevier.

studied circuits. The cellular phenomena described in locusts and chicks, as well as in other species such as the conveniently translucent zebrafish embryo, are finally getting a satisfying molecular explanation.

Pathfinding involves pairs of receptors and ligands situated in strategic, graded positions along axonal pathways. These pairs guide the leading tips of outgrowing neurons by attracting them toward some areas and repulsing them from others. There is now a broad understanding of how external guidance cues function to show neurons the way. Some reside on the surface of cells along an axon's path; others are secreted in concentration gradients and thus can act over longer distances. Scientists are also developing a grasp of the details of how a growing neuron processes this information internally to ensure it can respond appropriately by changing its gene expression to rebuild its inner skeleton, move, and finally stop to set up synapses.

Three genetic systems have predominantly informed this molecular study: the fruit fly, the roundworm, and the mouse. In recent years, the zebrafish has become increasingly useful as well, after an understanding of its genetics was developed in its early days partly with McKnight support to Monte Westerfield at the University of Oregon (Investigator, 1991). Scientists are now switching back and forth easily between simpler models and mice. They exploit the fly's and the worm's versatile molecular genetics to discover new genes and then analyze their molecular context in the more relevant mouse.

> *Throughout the study of neural development, McKnight has continued funding the people who were making the next steps in the larger development of the field.*

Scientists are alternately using genetics and biochemistry to unravel which genes and proteins control how large a synapse grows, how its pre- and postsynaptic sides communicate with one another as the synapse matures, and how much neurotransmitter the synapse needs to release. In this last area, Richard Scheller, who received a Scholar Award in 1984, has since become a leader through his discovery of many proteins that are critical for the packaging and release of transmitter-filled vesicles from synapses.

Throughout the course of this study of neural development, McKnight has continued funding the people who were making the next steps in the larger development of the field. Coming from Goodman's lab, they include Michael Bastiani, Jonathan Raper, Kai Zinn, Nipam Patel, Alex Kolodkin, David Van Vactor, and Aaron DiAntonio. Coming from Jessell's, they include Tessier-Lavigne, Samuel Pfaff, and Henk Roelink.

McKnight support for developmental neuroscience reaches far beyond these two lineages. Indeed, a large proportion of accomplished U.S. researchers in this area have received support early on in their careers, when they needed freedom to make the discoveries that subsequently built their reputation. They include Ben

Barres, Jane Dodd, Gregor Eichele, Scott Fraser, Frank Gertler, William Harris, Mary Hatten, Christine Holt, Jeff Lichtman, Liquin Luo, Susan McConnell, Peter Mombaerts, Dennis O'Leary, Barbara Ranscht, Scheller, Shatz, Larry Zipursky, and others. Other important contributors to developmental neuroscience received McKnight support later in their careers, when they needed funds to change the direction of their work. Examples include Gerald Fischbach, Jack McMahan, and Joshua Sanes.

THE VISUAL SYSTEM

The development and function of the visual system are other areas where the McKnight program has nurtured the growth of a large body of knowledge about the brain. It showcases the relevance of developmental processes to adult function, and it forms the foundation for the efforts of current neuroscience to begin to understand some of our cognitive processes, such as attention or the perception of context and meaning. With contributions from many McKnight-associated investigators, the mammalian visual system has become, to date, the best-understood example of how the brain establishes the astonishingly intricate wiring pattern of its adult form and of how it uses sensory experience to change its circuits continually.

Overall, scientists have painted a detailed picture of an embryonic brain that initially uses intrinsic cues to establish a basic blueprint of neural wiring from the eye to the visual cortex in the back of the head. In so doing, the emerging brain relies on chemical guideposts, many of which are identical to the ones that function in the developing spinal cord and were identified by Goodman, Jessell, and their trainees and colleagues.

> *With contributions from many McKnight-associated investigators, the mammalian visual system has become the best-understood example of how the brain establishes the intricate wiring pattern of its adult form and of how it uses sensory experience to change its circuits continually.*

Carla Shatz, professor and chair of neurobiology at Harvard Medical School and current president of the EFN, has led advances and established a lineage of younger investigators with her research on the next phase of visual-system development, which involves activity-dependent tuning. Starting in the womb, the immature visual system sharpens an initial, imprecise pattern of connectivity by strengthening appropriately active synapses and pruning excess ones. This process begins as the embryonic visual system rouses itself by means of spontaneously generated neuronal activity.

Activity driven by outside stimuli takes over after birth, once the newborn opens its eyes and the brain begins responding to sensory experiences from its surroundings. In this way, intrinsic "brain waves" and, later on, external input shape an initially approximate configuration of neural connections into its precise, mature form. Throughout life, as well, neuronal activity driven by a person's experiences strengthens some and weakens other synapses in a process of learning. This field already has helped to explain the development of children raised in neglect and sensory deprivation, as well as to correct certain problems with children's eyesight.

The story begins at Harvard Medical School where, in the 1960s, David Hubel and Torsten Wiesel (who served as president of The McKnight Endowment Fund in 1998 and 1999) used electrode recordings from single neurons in cats and monkeys to study the organization of the cerebral cortex. They found cortical regions where individual neurons were tuned to stimuli coming from a particular orientation, while others were selective for images sent in through either the right or the left eye. Hubel and Wiesel found that the cortex is organized into columns, vertical clusters of cells that span its layers. They introduced the concept that all the neurons in a column, from the cortical surface on down toward its deeper layers, share the same tuning (i.e., they preferably react to the same type of stimuli). There were columns for orientation selectivity. There were columns for eye selectivity, alternating millimeter-widths of cortex that correspond preferentially to the left or the right eye. Called ocular dominance columns, they have become a touchstone of higher brain structure in the years since.

Shatz and Michael Stryker are among Hubel and Wiesel's trainees who have developed the field from their mentors' initial discoveries. In 1978, for example, when Shatz and Stryker were still postdoctoral fellows, they described how these ocular dominance columns first form in a diffuse pattern whereby alternating groups of incoming fibers from the right and left eye overlap. Then, in a tremendous show of plasticity by the immature brain, fibers from the right and the left eye separate into segregated stripes during the first few months after birth. In fact, experiments preventing one eye from seeing for a while led that eye's ocular dominance columns to shrink, while the seeing eye's columns widened.

Hubel and Wiesel's pioneering work provided a vivid demonstration that changes in sensory experience can change the very structure of the brain itself. The window of time in an animal's life during which the columns could shift in this way was called the *sensitive period*. Its appreciation influenced the medical treatment of children with lazy eye and congenital cataracts, whereby the healthy eye is patched temporarily to shrink its cortical representation and give the impaired eye a chance to compete for more brain territory of its own.

Later, Stryker discovered that the ocular dominance columns begin to form even before the animal opens its eyes. Seeing sharpens the columns and keeps them malleable throughout the sensitive period after birth, but most scientists now think that spontaneous neural firing jumpstarts the formation of the ocular dominance columns *in utero*.

To understand better how activity sculpts the visual circuit, Shatz and other researchers harked back one step in its anatomical wiring diagram and focused not on the cortex but on a relay station deeper inside the brain. Called the lateral geniculate nucleus (LGN), it receives axons and nerve terminals from the retina, processes incoming signals, and then passes on its own fibers in the visual cortex. In this relay station, as well, an ordered pattern of eye-specific segregation forms even in the dark of the womb, whereby input from the left and the right eye gets sorted into characteristic alternating layers. Shatz and colleagues found that these layers in the LGN form in response to early spontaneous activity generated in the retina. Similar to what happens in the ocular dominance columns later, activity helps the axons coming from each eye to extricate themselves from an initially overlapping pattern to one where the input from the two eyes becomes neatly segregated.[4]

What exactly is this "blind" activity emanating from the retina? Work by Italian researchers showed that fetal neurons in the immature retina fire spontaneous bursts of action potentials. Shatz and her trainees showed that the bursts are coordinated into waves, which travel to the LGN and begin to fine-tune connecting neurons there. This last finding developed initially from the clever use of electrophysiology in cultured eye-brain preparations. Using confocal microscopy, researchers then observed spontaneous waves of calcium signals sweeping across the retina in real time. The definitive description, however, came by way of multielectrode recordings, a new technology that makes it possible to monitor the activity of many neurons simultaneously. Multielectrode recording has come into wide use for studies of brain function conducted at the systems level. It is being made ever more sophisticated by Markus Meister, who received a 2006 McKnight Technological Innovations in Neuroscience Award, and by McKnight awardee Eduardo Chichilnisky (Scholar, 2000; Technological Innovations, 2004), who is exploring how groups of neurons process visual information in the adult retina.

McKnight awardee Marla Feller (Scholar, 2002), a physicist-turned-neuroscientist who sometimes collaborates with Chichilnisky, has advanced the biophysical and molecular study of these waves, exploring how they encode information for the refinement of brain circuitry that has long been postulated. While training with Shatz, she found that amacrine interneurons in the retina trigger the waves in the retinal neurons that project into the brain and that certain

---

[4] The LGN layers and the ocular dominance columns are not to be confused with a different sort of visual map—that by which topographic information from across the eye's visual field gets correctly represented in the brain. On this question of temporal-nasal/up-down mapping, early pioneering work proposed that chemical cues might be at work; indeed, molecular studies later identified proteins called ephrins and Eph receptors. Perhaps the most prominent advances in this area came from Friedrich Bonhoeffer's group in Germany, but McKnight awardee Dennis O'Leary contributed important findings as well.

frequencies are required for the waves to be effective. Pushing this analysis to the molecular level, Feller has found that the rhythmic pattern of the waves mirrors oscillations in the activity of certain intracellular chains of signaling molecules, notably calcium and cAMP (see more on cAMP on page 65.)

Spontaneous waves of neuronal action potentials soon turned out to emanate not only from the immature retina but also from other developing areas of the central nervous system, including the hippocampus, an area known to be necessary for memory formation, and the spinal cord, where McKnight awardee Lynn Landmesser studies them. Like many fundamental discoveries made by McKnight researchers, spontaneous waves have become a general paradigm of how the nervous system can begin to refine its initially imprecise circuitry before sensory experience from the outside world becomes available.

The retina is where visual information sets out as it travels through our brains. With regard to the end of its path, in the cerebral cortex, researchers have made tremendous progress since Hubel and Wiesel's days in tracing and understanding where visual information goes after it leaves the primary visual cortex at the very back of the head. The anatomical connections between multiple brain areas that handle the higher-level processing and interpretation of vision have become much clearer. The present-day challenge of parsing the mechanisms by which that processing works has fallen to systems physiologists.

This brief synopsis has highlighted but a few selected findings from a much broader ongoing program of study. The McKnight Endowment Fund has supported a range of investigators in the field of visual system development after the first enabling discoveries were made. This includes scientists who trained with Shatz, such as Richard Mooney (Scholar, 1994), Meister, and Feller; collaborators such as Stephen Smith (Technological Innovations, 2004) and Denis Baylor (Senior Investigator, 1994); as well as investigators such as Lawrence Katz (Investigator, 1994), Dennis O'Leary (Scholar, 1987; Investigator, 2000), and Michael Weliky (Scholar, 1999). These scientists made many of the discoveries summarized here. Additional McKnight contributors to the area of visual system development include Mriganka Sur (Investigator, 1988), Hollis Cline (Scholar, 1991), and Larry Zipursky (Scholar, 1986; Investigator, 1991).

SYSTEMS NEUROSCIENCE

The knowledge gained about the visual system's development forms a platform for the next wave of investigation, namely that of how our brain fashions a conscious recognition out of the visual information that comes streaming in through our eyes—how we actually know what we see.

For a funder with a mandate to nurture research on plasticity, memory, and its diseases, channeling a significant fraction of its available funds toward studies

A 2002 McKnight Scholar, Bharati Jagadeesh studies how primates recognize and distinguish people, places, and other things they see. Here, she recorded from neurons of the inferotemporal cortex of monkeys while they viewed realistic images, and then compared the similarity of the images to the similarity of the neurons' responses. Reprinted from the *Journal of Neurophysiology*, 94; Allred, Liu, and Jagadeesh, "Selectivity of Inferior Temporal Neurons for Realistic Pictures Predicted by Algorithms for Image Database Navigation," 4068–4081, © 2005; used with permission.

in neural development, as McKnight did for many years, required a leap of faith. The foundation supported the award selection committees in their hope that knowledge about the mechanisms by which the brain wires itself up during development would also yield insight about synapses and memory in the adult and aging brain. And indeed, it has turned out that many of the molecules and

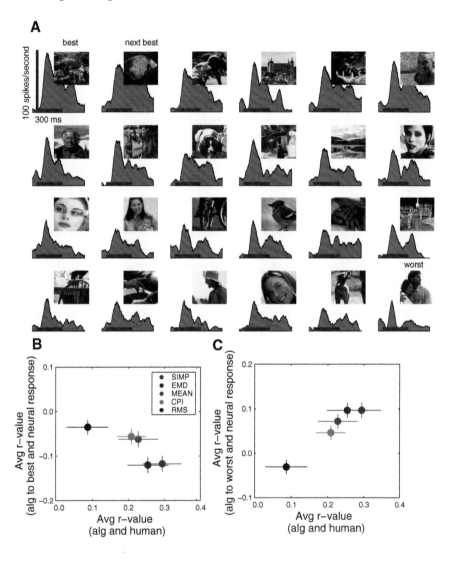

principles the brain and its synapses use throughout adult life begin in development.

To hark back to scientists cited early on, both Rutishauser and Landmesser belong to a growing group of developmental neuroscientists who for this reason have recently turned their attention to synaptic plasticity in the adult. (In 1997, Landmesser received a Senior Award to enable this change in direction of her research.) Some neurodevelopmental disorders and learning disabilities will one day be understood as disruptions in how precise circuits form. With regard to therapy development, it is worth noting that some of the molecules found to guide growing axons in the developing embryo have become experimental candidates in the preclinical study of repairing the injured spinal cord (see also "A Recipe for Success," page 79).

The visual system serves as a great object of study for much larger questions than development per se: how our brain fashions a conscious recognition out of the visual information that comes streaming in through our eyes—how we actually know what we see. Many labs are involved in systems and computational studies of visual processing. In this area as well, the McKnight program has engaged the leading minds since its inception and has made it a deliberate focus of its efforts to spot and support the most innovative thinkers so they will pull the field forward.

Systems neuroscience concerns itself with how the component parts of a given brain system are working together to produce a behavior. A central question in studying vision, hearing, and indeed all the senses is how a set of neural circuits processes sensory input into an appropriate motor output. This area underwent an early revolution when Edward Evarts and Robert Wurtz pioneered methods for obtaining recordings from single neurons in awake, behaving monkeys in the 1960s.[5] Before that, most investigators had been content to record properties of anesthetized animals. This transition to extracting information from awake animals instantly made results much more relevant to behavior. It is a revolution in neuroscience that happened in parallel to the field's adoption of molecular biology and, as systems neuroscience matures, may come to be seen as equally important.

For example, it opened up the study of attention and the eye's movements tracking an object. Unsurprisingly in hindsight, scientists found that the electric activity of a given neuron perked up when the animal's eye moved to follow the stimulus to which that neuron would tend to respond. Thus, scientists began to link the firing patterns of a neuron with perception. In this area, Stephen Lisberger (Scholar, 1981) is an example of someone who received McKnight funds early in his career and established a field that has become a mainstay of neuroscience. Likewise, McKnight awardees William Newsome (Investigator, 1988; Technological Innovations, 2000) and Richard Andersen (Scholar, 1983; Technological Innovations, 2000; Brain Disorders, 2005) have led the study of related questions about how the brain organizes behavioral responses to visual information.

---

[5] Evarts, of the NIH, was an early consultant and committee member for the McKnight neuroscience program. Wurtz was at the National Eye Institute, part of the NIH, and served McKnight as a committee member for the Senior/Investigator Awards in the late 1980s and for the Technology Innovations Awards more recently.

Other McKnight scientists advanced the field by studying perception and action. They have looked at how the brain extracts meaning from a given stimulus and how it stores information about its context. With a 2002 McKnight Scholar Award, Bharati Jagadeesh explored the question of how the inferior temporal cortex of macaque monkeys filters out irrelevant background information when the animal recognizes an image. Meaning and context can greatly change how an animal perceives a given image and cause corresponding changes in the neuronal response to the image.

The ability to study separately the different aspects of a given stimulus — its form, context, color, and motion — is now enabling scientists to tackle the so-called binding problem in cognitive neuroscience, which asks how the different brain areas that handle the separate aspects of a stimulus integrate their own respective neural activities into a unified perception. What is it about the neurons' firing pattern that encodes this kind of information? Innovators such as Brian Wandell (Senior Investigator, 1997), Michael Shadlen (Scholar, 1995), Richard Krauzlis (Scholar, 2000; Technological Innovations, 2006), Jonathan Victor (Scholar, 1984), and others have greatly increased the sophistication with which these complex questions have been tackled since the 1980s. Together, this kind of work, particularly findings by Charles Gilbert (Investigator, 1991), has thrown out the commonly held notion that the cortex is fixed into a rigid pattern of

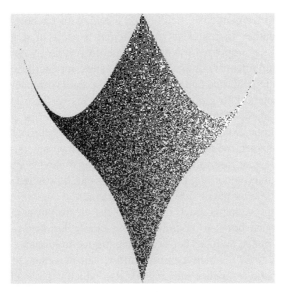

In looking at how neurons represent and process information, Jonathan D. Victor (Scholar, 1984) portrays the statistical limits of human vision in a two-dimensional "texture space." Courtesy of Jonathan D. Victor.

connections once it has matured. Instead, the cortex even of an adult retains an awesome capacity for reshaping its connections in response to experience.

Systems neuroscience has spread far beyond vision. For example, motor and somatosensory systems have revealed strikingly similar principles of cortical representation of the outside world. The primary motor cortex contains columns of neurons dedicated to analyzing inputs of given joints; the somatosensory cortex features its own topographic reproduction of the outside world. As another example, scientists are now tackling problems that are uniquely relevant to human behavior. Earl Miller (Scholar, 1996) studies how specific neurons in the prefrontal cortex encode rules about certain tasks. Paul Glimcher (Scholar, 1996; Technological Innovations, 1999) investigates the role of the parietal cortex in making decisions and judging the desirability of actions. Nikos Logothetis (Investigator, 1994) has identified neurons involved in categorizing objects. Other scientists study aspects of mood, such as elation and depression, with the same basic approach of trying to explain a behavior in terms of its underlying neural functions. Insights from psychology and sociobiology also inform these modern studies.

What has made this new level of analysis possible? The answer lies in large part with the availability of new technologies.

What has made this new level of analysis possible? As often, the answer lies in large part with the availability of new technologies. Multielectrode recording arrays have been tremendously helpful for listening in on how multiple brain systems fire together to represent the different aspects of a given stimulus. Michael Weliky has improved this technique even further by devising an array that records simultaneously from 64 different sites across the cortical surface without needing to penetrate and damage the brain. Paul Glimcher has developed an ultrasound method to guide recording electrodes to the right brain areas. Partha Mitra trained in theoretical physics and has used his 2000 McKnight Technological Innovations Award to learn experimental neuroscience so he can study implantable electrodes with an eye toward future prosthetic devices. David Redish (Technological Innovations, 2002) is working to devise a wireless remote control for implanted recording arrays so animals can be studied in their natural behavior without being handled by the investigator.

Technical improvements reach far beyond the electrophysiology equipment. Perhaps the biggest boost to cognitive neuroscience has come from methods enabling one to image cultured cells or whole brains. In humans, positron emission tomography (PET), which measures differences in regional blood flow, and functional magnetic resonance imaging (fMRI), which exploits changes in local oxygen levels following neural activity, make it possible to see brain areas in action while a person is performing a task or remembering past events. McKnight awardees are improving the applications of these technologies to neuroscience.

Seong-Gi Kim (Technological Innovations, 2001) and David Heeger (Technological Innovations, 2003) are using different approaches toward their common goal of increasing the resolution of fMRI such that it can visualize ocular dominance columns in the live brain. Nikos Logothetis is working to relate typical fMRI data more closely to neural activity by doing single-cell recording side by side with fMRI in awake monkeys. Scott Small (Brain Disorders, 2003) has improved a method for running MRI on mouse models to monitor the effect of potential treatments, and Dan Turnbull (Technological Innovations, 2000) has developed a way to visualize plaque pathology in a mouse model of Alzheimer's disease.

In animals, scientists have pushed toward a deeper level of resolution than is possible in humans by combining imaging with invasive molecular approaches. These include infecting neurons with viruses that carry fluorescent markers while also manipulating neuronal functions of interest. Such interventions can trace particular circuits, neuronal lineages, and the precise site of residence and action of particular proteins in the neuron. Many McKnight awardees have contributed to this rapidly advancing area; examples include William Harris (Scholar, 1980; Senior Investigator, 1994), Scott Fraser (Scholar, 1984), Christine Holt (Scholar, 1986), Jeff Lichtman (Investigator, 1985; Technological Innovations, 2005), and Joshua Sanes (Senior Investigator, 1997).

Since the late 1990s, the directors of the McKnight program have begun to shift the emphasis of the basic science part of the program gradually away from a preponderance of molecular studies and toward a greater balance with systems studies. "We did that because by then there had been many advances in molecular approaches," Gerald Fischbach said. "The reason we are interested in them in the first place is not a love of molecules per se. It's not even to explain how one nerve cell works, but how ensembles of nerve cells work in concert to produce brain function. That's the goal, and to do that we needed more analysis of circuits and systems."

This shift in priorities reflects the very success of the program. While important and knotty problems remain to be discovered in molecular neuroscience, the paradigms, the methods, and a basic plan for how to go about it are laid down. Therefore, some of the most adventurous among the coming generation of scientists are increasingly lured by the great question of how the mind emerges from the brain.

> *"Bridging the gap between what we can do in genetically accessible animals and what we want to know in humans will be one of the main challenges for the next few years,"* Tom Jessell said.

They have their work cut out for them. There remains a large gap between molecular and systems neuroscience. It is not yet clear what will be necessary to merge these two disciplines into a unified, molecular language for cognitive processes. A rich stream of genetics and molecular data continues to flow from

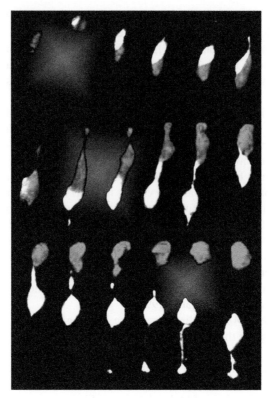

Like many McKnight awardees, William Harris (Scholar, 1980; Investigator, 1988; Senior Investigator, 1994) focuses on neurogenesis. These movie frames show how a cell divides and its daughters differentiate inside the retina of a living zebrafish embryo. Courtesy of William Harris.

fly, fish, mouse, and some other animals. By contrast, deep cognitive questions like language and consciousness—even the seemingly less imposing ones of sensory-motor integration—ultimately require that they be studied in humans and modeled in primates.

Some forms of human and primate genetics are available, but for ethical reasons it will remain largely sequence information and experimental data gleaned from cell culture, not the transgenetics that enable manipulation of brain processes. "Bridging the gap between what we can do in genetically accessible animals and what we want to know in humans will be one of the main challenges for the next few years," Tom Jessell said. "Even so, people have become emboldened to tackle the big problems."

Memory, and Losing It

The Fund's broad support of neuroscience reflects a guiding principle that important insights for memory loss will come from unexpected corners of science.

One of the most challenging problems is the one that inspired the backers of the McKnight neuroscience program from the outset: the workings of memory and its diseases. The McKnight program has always funded neuroscience broadly, and indeed has even structured its new Brain Disorders Award to include work on a wide range of neurodevelopmental and neurodegenerative disorders, as well as traumatic brain injuries. Nevertheless, its abiding interest has always been an understanding of memory and memory loss. The fund's broad support of neuroscience reflects a guiding principle that important insights for memory loss will come from unexpected corners of science.

How then has the study of memory progressed? A true unification of molecular and cognitive neuroscience has not yet occurred for memory or for any other faculty of the mind. Yet many of the component pieces for such unification have fallen into place. The anatomy of memory, its physiology, and a molecular science of synaptic plasticity—the cellular underpinning of learning—each have advanced impressively, and optimists dare say that with them, the field is poised for a functional integration and a big leap.

Consider first the anatomy of memory. It benefited from human and primate studies that influenced each other. Up until the 1950s, scientists believed that memories were stored all across the brain's surface, not in local, specialized subregions. Then in 1957, Brenda Milner caused a tectonic shift with her description of patient H.M., who, after surgeons removed both his medial temporal lobes to treat his epilepsy, became famously incapable of remembering anything new for longer than a few minutes. His intelligence, perception, and other cognitive abilities stayed intact, however, and even his childhood memories were still there. This drove home several fundamental points: (1) acquiring new memories is handled separately from the rest of cognition; (2) this process resides in the medial temporal lobe of the brain; (3) short-term (or immediate) memory, like repeating back a telephone number, occurs somewhere else; and (4) remote memories are stored elsewhere.

This experience yielded a rich vein of investigation. It drew many laboratories into delineating the different kinds of memory—for example, conscious and unconscious memories—and their respective sites of residence in the brain. Similar to what proved true in the study of vision, it became clear that with regard to memory, too, different specialized subareas of the brain handle its different aspects. Indeed, memory is composed of a set of brain systems, each with its own anatomy. It is still true that memory storage is widely distributed over the neocortex, but not in a random, general way. Instead, our brains have specialized equipment that is localized, which helps us store new memories. The storage of a whole event lies distributed in the different specialized subareas of the neocortex that each handled the processing of particular aspects of the event. Those different brain areas are bound together in part by the hippocampus and related structures in the medial temporal lobe.

EFN board member Larry Squire has had a leading role in this work. Influenced in part by H.M., in 1980 he established a monkey model for human memory impairment in his lab at the University of California, San Diego. Combining this model and the analyses of many human patients, his group subsequently identified the areas in the medial temporal lobe that host conscious memory, particularly the hippocampus, the entorhinal cortex, and the perirhinal cortex. The entorhinal cortex, which sends axons to the hippocampus, and a part of the hippocampus itself have since become widely accepted as the parts of the brain that are the first to malfunction when a person develops Alzheimer's disease.

Researchers have since learned that the hippocampus is a key structure for the brain's ability to fashion a lasting memory out of something it has just learned. This fits with the clinical appearance of early Alzheimer's, where a patient will remember newly learned information, say a name and address, for a few minutes, but 20 minutes later will have forgotten it completely, with no recall even if reminded.

The question of whether the hippocampus is truly important for memory formation has seen heated controversy for many years. Debate died down after 1986, when Squire published a report on the brain anatomy and histology of patient R.B. That patient had experienced memory impairment even though his brain damage was limited to the CA1 field in the hippocampus. Squire received a Senior Investigator Award (1988, 1991) to further his research on memory.

> *Scientists suspect that the human versions of learning genes will prove to play a role in certain mental illnesses, and The McKnight Endowment Fund is supporting early work in this direction.*

Other highly regarded memory scientists who received McKnight support at critical times in their careers include James McGaugh (Senior Investigator, 1977) and Richard Thompson (Senior Investigator, 1985, 1988). Young investigators include Anthony Wagner (Scholar, 2001), who has increased the time sensitivity of fMRI by combining it with magneto-encephalography. This allows Wagner to detect fast changes between the hippocampus and medial temporal lobe areas. In this way, he can pick apart how contextual information is packaged along with the original stimulus to build a complete memory and is then retrieved when the original stimulus reappears — or not, as all of us know who have encountered a familiar face and embarrassedly racked our brain for that person's name and for when and where we had met before.

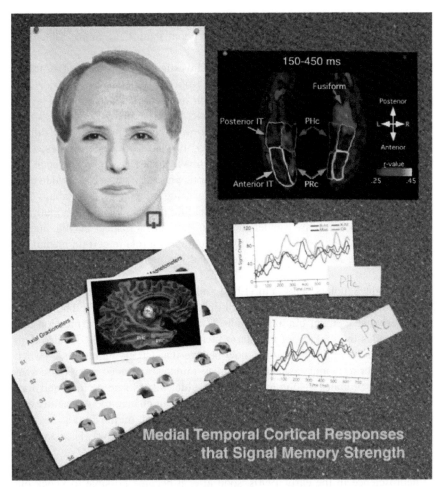

By combining fMRI and magneto-encephalography, Anthony Wagner (Scholar, 2001) can show how declarative memory enables a person to recognize previously encountered stimuli, such as a recently seen face. He focuses on the medial temporal lobe, a brain area that degenerates in Alzheimer's disease. Parts of the collage are reprinted from *Neuron*, 47; Gonsalves, Kahn, Curran, Norman, and Wagner, "Memory Strength and Repetition Suppression: Multimodal Imaging of Medial Temporal Cortical Contributions to Recognition," 751–761, © 2005, with permission from Elsevier. Artist: Anna Gitelson-Kahn.

What are these memories when viewed from the level of a cell? They are a collective set of synaptic changes in different brain areas. The capacity of a neuron to transform incoming information from the environment into changes in synapses is called synaptic plasticity. Ever since Kandel laid the groundwork for studying its molecular science and Seymour Benzer (Senior Investigator,

1991, 1994; Brain Disorders, 2004) opened up its genetics, this area of neuroscience has burgeoned in part thanks to the contribution of many McKnight awardees.

For example, Ron Davis (Scholar, 1984; Investigator, 1988; Brain Disorders, 2003) used McKnight support to expand from fly genetics to behavior and later to mouse studies, as a result of which he identified a number of genes involved in learning. He rescued a genetic learning deficit in a fly strain called Dunce by letting the hapless flies grow up and then slipping in the rat version of the Dunce gene. Published in 1995, this was the first genetic rescue of memory impairment in an animal, and it showed again just how universal the fundamental principles and proteins of synaptic plasticity are across a wide evolutionary range. Dunce, as well as other learning genes called Rutabaga and DCO, all fit somewhere into the cAMP signaling pathway of memory outlined only a few years before by Kandel and others.[6] Scientists suspect that the human versions of these genes will eventually prove to play a role in certain mental illnesses, and The McKnight Endowment Fund is supporting early work in this direction.

McKnight Scholars Jerry Yin (1996) and Tim Tully (1987) identified additional genes needed for learning in fruit flies. They also have studied how cyclic AMP response element binding protein (CREB) mediates gene expression that then precedes the making of long-term memories in the flies. CREB also does this in snails and mice. Scientists now believe that CREB represents a switch from short-term memory, which lasts many minutes and does not require synthesis of new proteins, to long-term memory, which takes longer to form and does require new protein to be made so that synapses can change their structure and function. These data validate early concepts studied in the 1960s by Sam Barondes, who demonstrated even then that protein synthesis was necessary for long-term but not for short-term memory.

The molecular exploration of synaptic plasticity expanded into a large subfield of neuroscience in the early 1990s, soon after it became possible to knock out specific genes in mice. Combined with pharmacology and electrophysiology of brain slices, such genetic experiments have identified specific sets of synapses and long-term changes in them that are necessary for specific memories. For example, the Schaffer collateral-CA1 synapse in the hippocampus is necessary for conscious memories of objects and space.

More broadly, these techniques have allowed a detailed analysis of long-term potentiation (LTP) in the years after Tim Bliss and Terje Lomo in 1973 first described this form of synaptic change in the hippocampus. McKnight-funded researchers have become leaders in the study of LTP and synaptic plasticity, including Robert Malenka (Scholar, 1990; Investigator, 1997), Chuck Stevens

---

[6] See the next section on ion pores for more detail on this pathway.

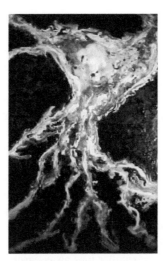

J. David Sweatt (Scholar, 1990) is an artist as well as a scientist. His painting expresses the complexity of cellular signaling in hippocampal pyramidal neurons, cells key to memory formation. Courtesy of J. David Sweatt.

(Senior Investigator, 1985), Steven Heinemann (Senior Investigator, 1994), Michael Mauk (Scholar, 1989), and J. David Sweatt (Scholar, 1990).

Since the early 2000s, methods of genetic engineering themselves have become more sophisticated. They now allow scientists to turn on or off their gene of interest at will at the time of their choosing, and only in particular tissues, and then examine LTP and other measures of synaptic function in the modified animal. This more specific manipulation has raised the experimental bar. Consequently, the thoughtful design of physiologically relevant animal models today has become a frontier in the exploration of both basic neuroscience and disease processes. Applied to the study of neurodevelopmental and neurodegenerative diseases, this is an area where EFN board member Huda Zoghbi[7] has established her reputation as a leader of the field.

ION PORES: A STORY LINE TO MEMORY, MENTAL RETARDATION

The roster of McKnight awardees since the 1970s includes numerous examples of how pioneering discoveries cross-fertilized each other and are beginning to

[7]Zoghbi's skill in devising mouse models followed her earlier discovery of the gene that, when mutated, causes Rett syndrome, an inherited form of mental retardation that affects primarily girls.

pay off in the form of mechanistic studies of brain disorders. A parallel arc worth spanning to illustrate this point follows the history of ion channels, which allow charged particles to pass through cell membranes according to different signals. By a different path, it also leads to synaptic plasticity, LTP, and molecules such as CREB. Ion channel research has yielded findings on epilepsy, pain (particularly migraine), demyelinating diseases such as multiple sclerosis, and other neurological conditions. Indeed, the term "channelopathy" has been coined to denote a growing group of nerve and muscle disorders. They can be genetic or autoimmune in origin but all compromise the proper functioning of ion channels.

A singular achievement in ion channel research was Roderick MacKinnon's (Scholar, 1992; Investigator, 1997) use of X-ray crystallography to obtain the first three-dimensional images of the potassium channel. Trained as a medical doctor, MacKinnon taught himself how to use this technique, and his inspired interpretation of X-ray data to explain how this channel manages to let large potassium ions pass while turning other, smaller ions away earned him a share of the 2003 Nobel Prize in chemistry. MacKinnon's pioneering work has since opened an active area of investigation into a range of other channels.

MacKinnon's research itself grew out of earlier work. In the 1970s, neuroscientists made headway in describing the biophysical and biochemical structure of ion channels. They isolated channels. They puzzled over how these pores distinguished one ion from another. They probed how they opened their gate to let ions flow through in response to the proper signal. Arthur Karlin, Bertil Hille, and William Catterall pushed this field forward. All three later received Senior Awards to move their research on ion channels to the molecular genetic level. (Hille's 1988 award was renewed in 1991; Karlin's award came in 1994; Catterall received his in 1997, and then in 2005 received a Brain Disorders Award to apply the research to the study of epilepsy.)

> *Voltage-gated and G-protein–coupled receptors opened a new door for investigations into what happens inside the neuron once it has received an outside stimulus.*

In the 1980s, two technical breakthroughs coincided to lay the groundwork for the molecular biology study of ion channels, setting the field on fast-forward. For one, German scientists Erwin Neher and Bert Sakmann invented an electrophysiological method that enabled scientists to measure the current flowing through a single channel.[8] For another, the advent of gene cloning made it possible to study the amino acid sequence of channels, modify them at will, and define how particular changes in the protein composition affected the channel's function. This kind of tinkering showed where the important parts of the

---

[8] It won these two German scientists a Nobel Prize in 1991.

channels — the voltage sensors, the gates — were on the overall protein. To everyone's surprise, these studies revealed that ion channels in the brain are remarkably similar to channel molecules throughout the animal kingdom and even plants and bacteria.

More to the point, these studies revealed that synapses function not just with channels that open and close directly in response to ions. They led to the discovery of another major class of channels, called the G-protein–coupled channels. These channels do not open right away to ion flows when the right ligand binds on the outside of the membrane, but open only after secondary signals on the inside of the membrane have been successfully touched off. The channels occur throughout the brain and, although they have a common basic structure, they come in a bewildering variety. Indeed, the discovery that a related family of more than a thousand different ones form a toolkit for our sense of smell earned 1992 McKnight Scholar Linda Buck the 2004 Nobel Prize for physiology and medicine, together with her former mentor Richard Axel.

Back in the early 1990s, voltage-gated and G-protein–coupled receptors opened new doors for investigations into what happens inside the neuron once it has received an outside stimulus. One outcome was a better understanding of long-term potentiation (LTP) and, more recently, long-term-depression (LTD). These forms of synaptic strengthening or weakening, respectively, together are thought to represent a cellular correlate of learning and memory. McKnight-funded studies of LTP have been conducted by Daniel Johnston (Investigator, 1982) and Robert Malenka (Scholar, 1990; Investigator, 1997). McKnight Scholar Daniel Feldman (2001) is currently working on LTD.

Crystal structure of the piece of the ion channel that is responsible for calcium-mediated control of the channel's activity. The first report of this structure came from the lab of Daniel Minor (Scholar, 2001; Technological Innovations, 2004). Courtesy of Daniel Minor.

Another outcome was that researchers began to pry apart what happens inside a neuron once it has received a stimulus from a neurotransmitter. Generally speaking, a G-protein–coupled receptor activates an ion channel indirectly through a third party called a coupling protein. This happens in interactions between proteins that frequently involve adding and removing phosphate groups from proteins with the help of enzymes. Many of the steps in this molecular communication process are classic biochemical reactions between proteins, cofactors, and enzymes, just like in the cells of any tissue in the body. The general finding that protein phosphorylation is a basic biochemical language inside neurons was developed in large part by McKnight awardee Paul Greengard (Senior Investigator, 1977, 1980), earning Greengard a share of the 2000 Nobel Prize. Eric Nestler, a 1989 McKnight Scholar and current EFN board member, has published widely in this area.

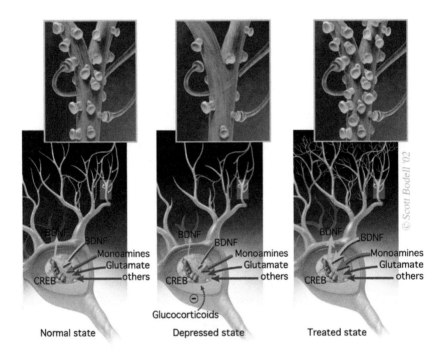

McKnight Scholar (1989) and current board member Eric Nestler examines neurotrophic mechanisms in depression. This image shows a normal neuron (left), the depressed state (middle), and amelioration by antidepressants (right). The neuron loses arborization, dendritic spines, and BDNF in the depressed state, particularly in response to stress hormones. Reprinted from *Neuron*, 34; Nestler, Barrot, DiLeone, Eisch, Gold, and Monteggia, "Neurobiology of Depression," 13–25, © 2002, with permission from Elsevier.

With the tools of molecular biology, investigators strung together chains of events that, at least in principle, bridge the gap from a stimulus to a rudimentary learning response in the neuron. It was already clear that long-term memory requires the making of new protein before it can be laid down in the brain. By contrast, short-term memory is thought to modify existing neuronal proteins. How, though, does the information that arrives at the synapse prompt the neuron to express genes in response?

With Greengard and other colleagues, Kandel teased out a bucket brigade of signal transduction steps in the sea slug that has since held up surprisingly well in mammals. In short, the pathway relies on cAMP, a small molecule that is a universal player in mediating how a neuron will react to a stimulus. cAMP sets a protein kinase on its way to travel to the nucleus. Once there, the enzyme prods other kinases into action, with the result that a gene regulator called CREB, now widely accepted as a key to long-term memory, turns on the production of certain genes. The proteins made from these genes then sustain the growth of new synaptic connections. Soon after, several investigators, including McKnight Scholars Alcino Silva (1995) and Mark Mayford (1997), found that cAMP, the kinase enzymes, and CREB are needed for LTP in the mammalian hippocampus, a site for forming our memories as well.

Granted, this initial path connecting a nerve impulse to a memory via a universal sequence of molecular steps is a first-pass approximation of all that goes on in a neuron during learning. Even so, it has awakened the interest of a great many scientists throughout neuroscience, who are now testing it, deepening it, and applying it to specific problems of learning and memory.

As an example from basic science, McKnight Scholars Kelsey Martin (2001) and Jerry Yin (1996) are wrestling with a new conundrum these discoveries have posed: If long-term synaptic change relies on newly made proteins, then how can this change be specific to the synapses where the stimulus came from? After all, gene expression is a cellwide phenomenon, but not all of a given neuron's hundreds or thousands of synapses ought to be strengthened just because one received an excitatory input. Martin has devised cell culture experiments with mouse neurons to show that there actually is lively communication between a remote synapse and the neuron's nucleus to make this process specific. The communication occurs in the form of traveling proteins that inform the nucleus of what goes on at its synapses.

Yin uses fruit flies to tease out exactly which proteins in turn tag newly made mRNAs to make sure they are active at only those synapses from which the signal originated. In one of the many fascinating cross-connections between different areas of biology, the proteins that appear to be controlling this address system are already known to function in other instances that require the uneven distribution of a cell's components. (One such example is the asymmetric division of stem cells. It yields another stem cell that retains certain cellular components so

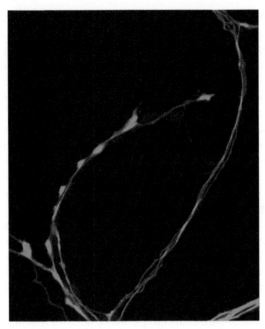

In this microscope image, Kelsey Martin (Scholar, 2001) shows cultured nerve cells from a sea slug. The sensory neuron (green) forms swellings, where the motor neuron (red) makes synaptic contact. A simple form of learning occurs at this contact. Courtesy of Kelsey Martin.

it can stay as is and keep dividing, as well as a daughter cell that gets a separate set of components on its way to maturation.)

In terms of memory, what happens once activity-induced mRNAs have made it to the right synapse? Martin, McKnight awardee Kimberly Huber, and other investigators such as Mark Bear, Oswald Stewart, and Paul Worley, began to fill this piece of the puzzle when they found over the past 8 years that certain mRNAs can be translated into protein locally right near particular synapses, not only in the cytoplasm surrounding the cell body, as was thought before.

How does this information apply to disease? For an answer, look to Fragile X, the most common form of inherited mental retardation. So called because of a distinctive nick in a chromosome of affected children, it is beginning to be seen as an example for what can go awry with a person's mind when the process of protein translation in synaptic plasticity is disturbed.

Children with this disease grow up with low IQ, attention deficit, and anxiety. They tend to show autistic as well as obsessive-compulsive behaviors, and many develop seizures. Their brains look normal at first glance, but actually their dendritic spines—tiny protrusions on which excitatory synaptic transmission occurs—are abnormally long, a sign that somehow synapses are not maturing properly. It is known that the disease is caused by a defect in the FMR1 gene that precludes the protein from being made.

Huber is an electrophysiologist by training. She turned her attention to this disabling condition when a disease model became available by deleting the gene from the repertoire of a mouse strain. This made it possible to explore exactly where things might go wrong with the mouse's synapses. For this goal, she received a McKnight Brain Disorders Award in 2002. Focusing on subtypes of the metabotropic glutamate receptor (glutamate is a major excitatory neurotransmitter in the brain), she found evidence suggesting that the FMR protein normally serves to rein in the local protein translation that occurs as a result of this receptor's excitation in the hippocampus and other brain areas.

The FMR protein resides just below synapses and associates with protein-generating organelles. The proteins made in response to activation of this type of the glutamate receptor are thought to implement many physiologic functions. Intriguingly, these functions overlap with the symptoms seen in patients. Specifically, these are their proneness to anxiety, seizures, and obsessive behaviors but also other symptoms having to do with a hypersensitivity to touch and even abnormal digestion. In this way, the FMR protein provides a nexus for studying the disparate symptoms of this mental illness at a molecular level.

One such function of the metabotropic glutamate receptor is LTD, the process of synaptic weakening that in newborn mammals normally helps eliminate excess synapses but is also thought to mediate a permanent loss of responsiveness when the baby does not receive appropriate stimulation. Daniel Feldman's McKnight support (2001) is for studies in this area. LTD works in concert with LTP, a mechanism of retaining those nascent synapses that receive the right kind of activity, and together the two sculpt neural connections as the brain matures in childhood. In Huber's experiments, the mice lacking FMR showed more clearly this LTD mechanism in their brain, suggesting that synapse maturation in this mouse is unbalanced and that the abnormal dendritic spines seen in postmortem samples of patients with Fragile X are actually weakened synapses en route to elimination. This could contribute to the cognitive problems in the affected children.

While this research continues, it already reaches far beyond prior clinical descriptions of mental retardation. What's more, the central thought that the disease is due to runaway responses to a particular subtype of glutamate receptor offers a potential future strategy for treatment. Antagonists to this receptor, while not replacing the lost role of FMR protein, could all the same tone down some of the receptor's unrestrained effects on the brain, and some experimental compounds already appear to do so. Besides raising hope for a future drug treatment against mental retardation, this study in a broader sense exemplifies how neuroscientists today reach beyond the techniques they trained in and combine genetic technologies, electrophysiology, and molecular biology to begin studying the mechanisms that cause mental illness.

Studies of Fragile X exemplify how neuroscientists today reach beyond the techniques they trained in and combine genetic technologies, electrophysiology, and molecular biology to begin studying the mechanisms that cause mental illness.

The McKnight program has produced many other leaders in neuroscience whose accomplishments are not covered comprehensively in this essay. To mention one last example, Solomon Snyder (Senior Investigator Award, 1977, 1980) is a pioneer of neurophysiology who in 1978 won one third of that year's Albert Lasker Basic Medical Research Award for his description of opiate receptors in the brain. In the years since, he discovered novel neurotransmitters such as nitric oxide and carbon monoxide, and in 2003 he received the National Medal of Science. His laboratory has made wide-ranging contributions to our knowledge of neurotransmission and helped lay a foundation for understanding the action of psychotropic drugs.

The roster of McKnight awardees reflects a breadth of topics and technical approaches, but for this particular period in the history of neuroscience, they all have one characteristic in common. "For all different approaches to brain function and disease," Jessell said, "the key has been to get to the molecular level of whatever the question is you are studying."

ALZHEIMER'S DISEASE: A CASE STUDY IN BENCH-TO-BEDSIDE PHILANTHROPY

The public at large, and surely any funder, may well ask where all this research leads. The answer has two parts. First, as investigators drill deeper into ever more complex realms of basic science, we learn about who we are as human beings. Science philanthropy fuels the continuing human quest for enlightenment, and without question The McKnight Foundation has contributed immeasurably to the goal of understanding our brain in healthy states. Second, the program's long-range goal remains, and always has been, to help us understand and treat diseases of memory and the mind. At this point in the program's history, it is too early to claim spectacular home runs for the diagnosis or treatment of neurologic and psychiatric diseases. There are, however, many triumphs at earlier stages in the journey from bench to bedside. It is this transition zone between basic science and new therapies where McKnight's funding choices have had an impact on the development of current experimental therapeutics. More broadly, the McKnight program has helped bring about a sense of optimism among the research community that neurologic and psychiatric diseases will become amenable to therapy within the next generation.

Generally speaking, academic science lays the groundwork for drug discovery and development, which then occurs in pharmaceutical and biotechnology companies.[9] Treatments and diagnostics are rarely traceable to a particular McKnight project. The foundation's impact lies more in its seeding of new lines of investigation, which then grow to a body of knowledge that industry can use. It is also true that lingering problems of length and costliness of clinical trials, as well as delivery of drugs to targets in the central nervous system, have kept the proportion of experimental drug candidates in neurologic diseases far below those for the nation's other top medical needs, such as cardiovascular disease, cancer, or diabetes.

Consider Alzheimer's disease (AD). It is a good case for illustrating the role of the McKnight neuroscience program in advancing the field, as well as for identifying the practical issues still holding back drug development. In AD, companies hesitate to launch clinical trials for experimental drugs in part because drugs tend to fail in patients with established disease, in whom the diagnosis is fairly clear. There is no quick and robust way to identify patients with the earliest signs of Alzheimer's, whom the drug at hand would be more likely to help.

Despite the hurdles still ahead, Alzheimer's disease showcases how profoundly experimental, molecular science has transformed our understanding of a baffling problem.

---

[9] Lately, these efforts have begun to overlap somewhat as academic drug-screening laboratories have sprung up as part of a broader push within academia to try to advance applied projects further before passing them on to a drug company.

Mild memory loss can have many causes, and enrolling people with causes other than incipient AD dilutes the trial population so that, again, the drug's effect may become undetectable. At the end of a trial, when quantifying the effect of the test drug, companies are reduced to using cumbersome batteries of neuropsychological and clinical assessments that show variance from trial site to trial site and require that trials run for 6 months before a reliable signal can be expected.

What's sorely needed is a reliable surrogate marker that flags right away whether a person is developing AD and whether a drug is working. A urine test as is used for pregnancy, a blood test as is used for signaling prostate cancer, or a one-time brain scan could, in theory, all serve this purpose.

A 2000 McKnight Technological Innovations Award to Dan Turnbull went in this direction. It supported his development of an MRI imaging agent to reveal the amyloid deposits that mark AD and that are known to begin building up years before people develop symptoms. Once validated, such an imaging marker could identify AD early and track the success or failure of experimental drugs designed to reduce the amyloid burden of the brain. A different amyloid imaging agent is being tested successfully in humans, but it relies on PET (positron emission tomography), and eventually an MRI agent will be necessary to make such a diagnostic scan widely available.

Despite the hurdles still ahead, Alzheimer's disease showcases how profoundly experimental molecular science has transformed our understanding of a baffling problem. "Alzheimer's research exemplifies the revolution of clinically relevant neuroscience in the last 30 years, and most of this was made possible by new tools and ideas that came from basic scientists," Barondes said. Axelrod recalled in a 2004 interview that the McKnight advisers avoided funding targeted or descriptive AD research, instead placing their bets on a broader strategy. They were right. The revolution in AD research occurred in large measure through biochemistry, genetics, cell biology, and imaging technology.

Progress on Alzheimer's has generated widespread optimism that some of the diagnostics and mechanism-based treatments currently wending their way through preclinical and clinical development will eventually pan out. They will come not a day too soon: demographers predict that up to 15 million Americans will suffer from this disease by 2050 unless better medicines are found. Cost estimates of current Alzheimer's care range widely but reach $100 billion annually; the emotional cost on families is incalculable.

What has led to this optimism? The short answer is that basic neuroscience has pieced together a solid picture of the pathogenic pathway that leads to AD. Though the ultimate cause of sporadic AD remains an enigma, this central pathway is thought to be at play in all cases. The main advantage of this discovery is that it offers many points of entry for drug developers to try to interrupt it. To be sure, the pathway still has holes, even active detractors. After all, trying to replicate or disprove a colleague's findings is part of the scientific method. That said, a majority of researchers in the field largely agree on how AD develops, and researchers in diseases where progress has been slower, such as schizophrenia, look to AD for encouragement of what lies ahead for their own fields.

SLOW PROGRESS, THEN A BREAKTHROUGH

The breakthrough came with biochemistry and genetics. Until the 1970s, researchers in the field had been able to do little more than describe in more detail Alois Alzheimer's original presentation, in November 1906, of the disease's two pathologic hallmarks to an audience of fellow psychiatrists in Tuebingen, Germany. The radical element of Alzheimer's lecture was that by identifying neurofibrillary tangles inside neurons and amyloid plaques in the spaces between neurons, he blamed physical changes in the brain for a disease that presented as loss of memory, mind, and sanity. But for a long time after that, research was limited. New work consisted of using the electron microscope for higher-resolution images of the plaques and tangles, and of finding them in other species, such as old rhesus monkeys. The leading AD pathologist at the time, Robert Terry, then at Albert Einstein College of Medicine in the Bronx, was involved in the McKnight neuroscience program early on as an adviser and received Senior Investigator Awards in 1981 and 1984.

One foretaste of the changes to come was the discovery, in 1969, that all people with Down's syndrome develop Alzheimer pathology by middle age. This hinted that chromosome 21, of which Down's children inherit an extra copy, might hold answers to what goes wrong in AD.

The scientists advising The McKnight Foundation at the start were watching these emerging developments closely. In fact, in a 1980 meeting with Virginia Binger, William McKnight's daughter who at the time presided over The McKnight Foundation, they noted senile dementia and Alzheimer's disease in Down's patients as among diseases on the verge of a cure. Though this assessment proved overly optimistic, it did reflect the excitement that gripped neuroscientists when the first mechanistic news about Alzheimer's disease began to break after decades of descriptive anatomy of postmortem tissue.

On a separate track, scientists in the late 1970s realized that certain types of neurons — those using acetylcholine as their transmitter molecule — were dying disproportionately in Alzheimer's disease. Peter Davies made this initial discovery in 1977 and soon would become a leading AD pathologist while working with Terry. Davies's observation gave rise to the cholinergic hypothesis, which states that age-related memory loss results from a deficit of cholinergic neurotransmission, much as Parkinson's disease is marked by a deficit of dopamine-producing neurons in a different brain area. McKnight Scholar Peter Whitehouse

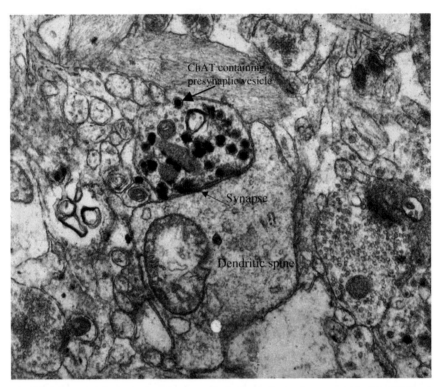

The first demonstration of a cholinergic synapse in the human brain, from research by Marek-Marsel Mesulam (Senior Investigator, 1985). Cholinergic neurons are especially vulnerable in Alzheimer's disease. Reprinted from *Experimental Neurology*, 144; Smiley, Morrell, and Mesulam, "Cholinergic Synapses in Human Cerebral Cortex: An Ultrastructural Study in Serial Sections," 361–368, © 1997, with permission from Elsevier.

(1982) helped formulate this hypothesis, and Marek-Marsel Mesulam (Senior Investigator, 1985) has described the anatomy of the cholinergic pathways affected in AD.

> *Today's drugs improve symptoms for 6 months, but they do not address the underlying mechanism of Alzheimer's disease.*

The hypothesis inspired the development, over the next decade, of today's batch of approved AD drugs, which increase acetylcholine levels around the remaining synapses to boost synapse function. (A more recent drug acts differently.) These drugs are in wide use, but even their makers acknowledge that their effect is modest and transient. They improve symptoms for up to a year, and that is valuable for patients and families while no better drugs exist, but they do not address the underlying mechanism of Alzheimer's disease.[10] In addition, other work showed that other brain regions, the entorhinal cortex and hippocampus, are especially vulnerable at the early stages of the disease.

That underlying mechanism began to come to light in the mid-1980s, when Californian scientists, followed a year later by German-Australian colleagues, isolated the small amyloid-beta protein that makes up the plaques. Another year later, in 1986, French, Japanese, and American groups independently identified the key component of the tangles as being the protein tau. One of these scientists, Dennis Selkoe, was funded in his postdoctoral years by a grant to his adviser, Michael Shelanski (Senior Investigator, 1977, 1980) and in 1988 received his own Senior Award. Also in the 1980s, Terry would use his McKnight Senior Award to support biochemical work by Davies, who became a leading expert on tau.

Biochemistry had burst on the scene. Scientists were no longer looking at AD tissue and staining it with dyes. They were grinding it up and separating its component proteins. The field underwent a first expansion, as many new groups seized the newly available proteins and probed them using newly available technologies of molecular cloning. They found that the amyloid-beta protein got clipped from a larger precursor protein called APP that lodged in the neuron's membrane and indeed occurred in most tissues of the body. The clipped piece could form fibrils in a test tube and was toxic to neurons. The tau protein in AD somehow abandoned its normal job of stabilizing the internal struts that make axons rigid and instead clumped together into toxic filaments in the neuron's cytoplasm, leaving the axons to crumble.

The field was also branching out into other aspects of the disease. An inflammatory component became apparent, and scientists began to point an accusing finger at the apolipoprotein E. Moreover, scientists realized that synapses malfunction in AD long before the whole neuron dies. They began to suspect that those troubling memory problems that give away early AD in an otherwise still-functioning person may actually reflect dying synapses as much as the loss of whole sets of neurons.

In 1987, human genetics of AD was gearing up. Its findings over the next few years would create a defining point of convergence in the understanding of this disease. Investigators began by following up on the clue provided by Down's patients and soon mapped the first AD gene to chromosome 21. In this effort, a 1985 McKnight Investigator grant to Harvard geneticist James Gusella and another in 1988 to Nikolaos Robakis at Mount Sinai had a great impact on AD research while these investigators were racing to pin down that gene.

---

[10] More recently, a second wave of interest in a new generation of experimental AD drugs has built up around targeting specific acetylcholine receptor subtypes that appear to influence the amyloid mechanism of AD.

Over the next few years, several groups tried feverishly to identify the gene with AD-causing mutations. In 1990, the results started coming in rapid succession. That year, Dutch/British/American researchers found mutations in the APP gene causing a rare form of an AD-like dementia. Other geneticists in 1991 found APP mutations in families with hereditary AD, followed by different mutations in 1992. To date, 25 mutations in APP are known. In 1993, the role of the ApoE gene was exposed: its E4 variant greatly increases a person's risk for developing AD and reduces the age at which the disease strikes. In 1995, researchers identified a third gene as mutated in familial forms of AD. Called presenilin, its two forms together today account for half of familial AD cases. Some people carrying such mutations develop the disease as early as their late 30s, leaving families and young children to watch as it obliterates their loved one's mind and spirit. Every year, additional mutations are added to the more than 165 already known as more families enter genetic studies. In 1997, scientists identified another protein-cutting enzyme, BACE, which precedes the action of presenilin and is now considered the first molecular step toward pathology. AD-causing mutations in tau were never found; however, the 20 pathologic tau mutations that are known to date cause related dementing disorders called tauopathies.

The convergence of all four known genes on the amyloid-beta protein was a rare, lucky break.

The transforming convergence that centered the field's attention squarely on the amyloid-beta peptide occurred in the late 1990s, when scientists realized that the presenilin genes encode the catalytic component of the long-sought enzyme that clips the pathogenic amyloid-beta peptide away from its normal precursor protein APP. It also became clear that all mutations known in either APP or the presenilin genes somehow tended to increase the load of amyloid-beta protein in the brain. The mechanism by which the ApoE4 variant accelerates AD is still inconclusive, but even there, the most widely accepted hypothesis holds that its protein does a poor job of clearing excess amyloid-beta peptide and instead accelerates its accumulation.

This convergence of arguably all four known genes on the amyloid-beta protein was a rare, lucky break. Typically, complex diseases that develop from mutations or variants in multiple genes tend to remain quite mysterious long after a few of those genes are first identified. Unlike the AD genes, the handful of genes known so far to play some role in schizophrenia, or Parkinson's for that matter, have not yet similarly encouraged scientists by all fitting into one common function. In the case of AD, the converging genetics and its associated biochemistry led to the formulation of the amyloid hypothesis by Selkoe and other investigators. It has been modified over the years, but its essence states that AD develops when accumulating amyloid-beta protein aggregates, which in turn promotes inflammation and the aggregation of tau, interfering with the function of synapses and eventually leading to the death of neurons.

The availability of the genes for study drew additional investigators to the field, and it expanded a second time. This expansion drew further fuel from genetic association studies that try to link additional genes to AD. None have turned up so far that are as solidly proven as APP, the presenilins, or ApoE, but several hundred candidates that have been suggested give research groups all over the world biological rationales to test further. The ApoE4 variant is a risk factor for the common form of AD, which affects the elderly, whereas mutations in APP and presenilins cause rare forms of familial, early-onset AD.

Scientists found other protein-cutting enzymes that also clip the amyloid-beta precursor protein. They studied the complicated biology of how these various proteins move through the cell and where they meet to make the amyloid-beta peptide. They understood that the precursor protein gets cut in a complex manner right in the middle of a lipid membrane, a heretofore unusual site of protein cleavage that necessitated the development of new biochemistry methods to study it adequately. This is still a challenge. Cell culture assays became tools to screen for compounds that lower amyloid beta levels.

As soon as scientists had identified the genes, they began making mouse models with these mutations. These mice mimic only some aspects of AD. Indeed, finding a true model for this multifaceted disease has been arduous, and only the most recent strains that use more sophisticated genetics have become more satisfactory. Despite their drawbacks, the mouse models have become indispensable in assessing the effect of chemical compounds that can decrease the amyloid-beta load. Only since the early 2000s have adequate tau models become available, and they are spurring drug screens in industry against aggregation of this protein as well. Other species, such as flies, have joined mice in recent years as tools for exploring mechanisms and screening for antiaggregation drugs, as in the work of 2001 McKnight Brain Disorders awardee Mel Feany. Amyloid-lowering compounds and vaccines are currently in preclinical development, and some have entered human trials, though there have been setbacks in the clinic and none have made it to market yet.

LOOKING AHEAD

In 2006, the Alzheimer's disease field has become mature in that researchers are addressing finer points of the amyloid hypothesis. Progress is reflected in how the questions are changing. To be sure, even the most ardent proponents of the amyloid hypothesis do not claim that it explains everything about AD; they merely insist that it offers the best entry points for therapy development of any explanation to date. To complete the picture, scientists are reaching beyond amyloid science to incorporate other processes that together will give a more encompassing explanation of the disease. The McKnight Endowment Fund supports innovation both within and outside the central hypothesis.

> *The McKnight Endowment Fund supports innovation both within and outside the central hyp0.5pothesis.*

First, consider recent work within the amyloid hypothesis. At Columbia, Scott Small (Brain Disorders, 2003) has combined technical advances in multiple component areas of neuroscience, such as MRI imaging, genomics, and cellular biology, to address a glaring weakness in our present understanding of AD. The majority of AD patients does not have mutations in the amyloid precursor proteins or the presenilins. How, then, could amyloid-beta levels rise in the many millions of sporadic cases worldwide? This open question keeps alive skepticism about the role of this peptide in the disease.

Small decided to approach this question with DNA microarrays, a technology that was invented in part by the founding chair of the McKnight Technological Innovations Award Committee, Lubert Stryer. In the past, Small used MRI imaging to help pinpoint at a fine level of resolution the subareas of the brain that are most vulnerable to the earliest defects in AD. His work confirmed them to be the entorhinal cortex and the CA1 subfield of the hippocampus, and he drew on this knowledge for the microarray study. He sampled just these tiny specks of human brain tissue for a genomics analysis of which genes are expressed differently in people with sporadic AD.

Small focused on one suspicious gene whose expression was high but protein levels were strangely low. This protein appears to be necessary for the proper transport of the BACE enzyme from one cellular compartment to another. The insufficient supply of this transport protein creates a backlog of BACE in little membranous sacs called early endosomes. There the lingering enzyme suddenly finds itself cheek by jowl with the APP protein and thus can cut it, which in healthy neurons it rarely does. This is a speculative scenario, but human genetics suggest that some people with late-onset AD have variations of this gene that might make them prone to the disease. This discovery is helping advance the current edge of AD cell biology, which is shining a renewed spotlight on the cellular transport routes that redistribute APP and its cleaving enzymes.

A 2005 Brain Disorders Award to the noted structural biologists Gregory Petsko and Dagmar Ringe at Brandeis University is enabling them to tackle a different problem that has long dogged Alzheimer's research. It concerns tau. Though undeniably a central component of AD pathology — and indeed numerous other neurodegenerative diseases — this protein has stubbornly resisted attempts at figuring out exactly what it does in the disease process. Tau has been subjected to thousands of studies, and its normal function is very well known, yet precisely how it becomes toxic to neurons in degenerative diseases remains mysterious. Petsko and Ringe, a husband-and-wife team, are taking a fresh approach to this problem by studying tau in yeast.

Finally, consider an example of the surprising new ideas that enrich our understanding of Alzheimer's from outside of established thinking. McKnight awardee Ben Barres is not an AD researcher. He is interested in the development of glial cells, the three-pronged support staff for the neurons. There are astrocytes, oligodendrocytes, and microglia. Each of these cell types is thought to play some role in AD, but collectively they receive not nearly as much attention as do neurons when it comes to studying neurodegeneration. Barres has twice used McKnight funding (Scholar, 1993; Investigator, 1997) to advance an idea that was too far outside conventional wisdom to receive support from more conservative funding agencies.

Barres wanted to follow up on his hunch that, rather than being passive support cells as most people thought, glial cells might quite actively support the formation of synapses in the brain. In fact, discussion with scientists at a McKnight conference strengthened his belief in the validity of his idea, even though federal funders rejected it at the time. McKnight enabled initial studies, and Barres found that astrocytes increased by a hundredfold the activity of synapses between neurons kept in cell culture.

When that was not enough to secure funding from the National Institutes of Health, McKnight stepped in again and funded follow-up work, which found that cultured neurons form many more synapses between them when astrocytes are present also. Since then, Barres has shown that this phenomenon is more than an artifact of cell culture but happens in the developing brain. What's more, he has identified an astrocyte protein that triggers the formation of synapses, called thrombospondin. Astrocytes secrete this protein around the time synapses form in the brain. Then they shut off its production in the adult brain, but turn it back on after injury. This protein has come up in AD research before, but so have many other proteins, and the field has not so far followed up in detail. Barres is now turning his attention to what role astrocyte proteins may play in the fate of synapses as the disease develops.

Chapter 3

A Recipe for Success

WHY McKNIGHT'S PROGRAM HAS MADE AN IMPACT

Young neuroscientists regard McKnight Awards as the most prestigious they could obtain.

—Liqun Luo

Given the conservative funding of government sources, the support obtained by foundations with the McKnight policy will be our hope for innovative research.

—Nikos Logothetis

By the year 2006, The McKnight Endowment Fund for Neuroscience was able to look back on a singularly successful 20 years of neuroscience funding and 30 years since the parent foundation first decided to support neuroscience. By almost any measure, The McKnight Foundation has established a stellar reputation for picking the best and the brightest at the most productive time in the person's career. Forty-nine of the roughly 125 neuroscientists who are currently members of the National Academy of Sciences have been McKnight fellows, most before their election to the academy. Similarly, 42 neuroscientists who are or have been Investigators of the Howard Hughes Medical Institute were McKnight fellows in the past. Leaders throughout all areas of modern neuroscience have come through the McKnight program.

McKnight awardees have received prestigious awards for work they did, in part, with McKnight funding. To cite only one, consider the Nobel Prize. From 1977 to 2006, there have been seven Nobel Prizes for neuroscience, which were

Research Funding in Neuroscience
79

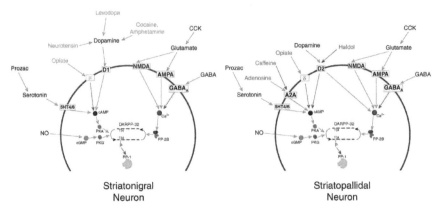

Striatonigral
Neuron

Striatopallidal
Neuron

Paul Greengard (Senior Investigator, 1977, 1980) was a co-recipient of the 2000 Nobel Prize for his work on how neurotransmitters act within the nervous system. This diagram illustrates how signaling pathways mediate effects of neurotransmitters and drugs in the two subtypes of striatal projection neurons, some of which degenerate early in Parkinson's disease. Reprinted with permission, from the *Annual Review of Pharmacology and Toxicology*, 44, © 2004 by Annual Reviews, www.annualreviews.org.

shared by 16 scientists. Half of them did their seminal work before the McKnight program began (including McKnight awardee David Hubel, who received the prize in 1981, 10 years before his McKnight Senior Investigator Award). Another two, Ernst Neher and Bert Sakmann, were ineligible for a McKnight Award because they worked in Germany. But of the remaining six, four had received McKnight funding: Paul Greengard and Eric Kandel (who each had a share of the 2000 Nobel Prize), Roderick MacKinnon (2003 Nobel Prize), and Linda Buck (2004). In one year alone, 1992, two of the six young scientists selected for a Scholar Award—MacKinnon and Buck—would go on to win the highest recognition in science. In all, five Nobel laureates received McKnight Awards, and two more, Julius Axelrod (1970) and Torsten Wiesel (1981, with Hubel), served the program as board or committee members to help identify the next generation of investigators.[1] (Truth be told, it's worth noting that one future Nobel laureate, Stanley Prusiner, slipped by the committee when he applied in 1985, a time when his research was still highly controversial.)

> *The program accelerated a thorough transformation of neuroscience from an anatomic-physiological and descriptive discipline to an experimental, molecular-biological one.*

[1] Another Nobel laureate who received McKnight funding was James Watson. However, his work on the double helix, for which he shared the 1962 Nobel Prize in physiology or medicine with Francis Crick, was unrelated to the work at Cold Spring Harbor Laboratory for which he received a McKnight Director's Award of $110,000 in 1984.

The influence of The McKnight Endowment Fund for Neuroscience (EFN) on the science itself has been both broad and deep. By choosing innovators and nurturing bold ideas at the very edge of current knowledge, the program in the 1980s accelerated a thorough transformation of neuroscience from a predominantly anatomic-physiological and descriptive discipline to an experimental, molecular-biological one. That revolution nearing completion, the EFN in the 1990s shifted its weight toward attempts to step back up from the level of molecules and use what we know about them for a renewed focus on how nerve circuits and whole systems function. Already looking ahead toward an integration of molecular and systems language, the EFN is poised to advance the leading edge of this next challenge with selected awards as well.

In the late 1990s, the EFN directors decided that a sufficiently solid stock of knowledge about fundamental neuroscience had accumulated in the past 25 years that it was time to call on it and start bridging the gap between basic studies and clinical problems. The directors reorganized the current award structure. They created new awards that enable basic scientists to turn their attention to mechanisms, diagnosis, and treatment of brain disorders, as well as to develop new technologies for addressing basic and applied questions.

What is it about the McKnight program that has made this success possible? After all, the amounts paid are small compared to federal research budgets. The program started out with $750,000 per year and now disburses about $4.4 million per year, whereas the two programs funded by the National Institutes of Health (NIH) that were most involved with neuroscience awarded more than $2.2 billion in 2004.[2] What is the model that has enabled this private funder to achieve an impact disproportionate to its size? The following 10 principles have guided the directors since the program's inception.

1. IDENTIFY EXCELLENCE

If I had to choose one word for the people on the board, on the review committees, and the awardees, it would be excellence, more than any other foundation I know. Its impact is the tradition of excellence it established.

—Gerald Fischbach

It's a who's who in neuroscience.

—Peter Mombaerts

[2] The National Institute of Mental Health gave $1.03 billion in grants in 2004, and the National Institute of Neurological Disorders and Strokes gave $1.2 billion. However, at least six NIH agencies give some of their research grants for neuroscience-related work, so the total annual federal expenditures on neuroscience research are likely to be much higher.

This principle clearly is the stock of the soup, and it was from the very beginning. "We strived for absolute excellence, because we knew that bad money drives out good," recalled Sam Barondes, a steward of the program since 1977. When people of the utmost regard make up the selection committees, they will attract the best applicants and maintain the standard for future committee and board members.

McKnight recruited excellence from all walks of scientific life and in turn created future leaders. Lineages of great scientists who went through the program reflect this. Examples abound in the McKnight record, but consider these two: The program began with the conceptual imprint of Julius Axelrod, a Nobel laureate who represented the great generation of NIH scientists from an era after World War II, when research at the NIH was a model for the country at large. Axelrod also embodied a pure spirit of meritocracy. One of his trainees, Solomon Snyder, is a prolific leader in neuroscience known for his discovery of opiate receptors, among many other achievements. Snyder in turn trained McKnight awardee David Bredt, who, while with Snyder, discovered that nitric oxide is a neurotransmitter and then quickly became one of the most highly cited young neuroscientists. Eleven years after his postdoctoral fellowship, Bredt assumed leadership of integrative biology research at Eli Lilly and Company.

A different lineage begins with Gerald Fischbach. In parallel to his research career in synapse development and physiology, Fischbach became a leader in science administration, heading, in turn, university neuroscience departments, the National Institute of Neurological Disorders and Stroke, and eventually the faculty of medicine at Columbia University. Fischbach trained Thomas Jessell, who is perhaps the foremost developmental neurobiologist of our time. Jessell trained Marc Tessier-Lavigne, who used McKnight money to make his first breakthrough in a series of discoveries about guidance proteins in the developing spinal cord. Tessier-Lavigne now applies his research to the development of drugs against cancer and other diseases at Genentech.

2. FIND YOUR NICHE

The McKnight programs are tremendously important in launching new, unproven investigators and new, unproven projects. By wise and adventurous choices, the McKnight committees have a huge impact on the course of discovery in neuroscience.

—Gary Yellen

The McKnight program has acted as an expertly targeted catalyst for promising ideas and scientists. It has enabled research that would otherwise have remained a "great idea" but never been explored.

—Eric Knudsen

These days, a variety of funding sources support neuroscience research. They include the six institutes of the NIH that fund neuroscience, the National Science

Eric Knudsen (Investigator, 1985; Senior Investigator, 1997) studies how the barn owl's exquisite hearing allows it to target its prey in the dark. In early life, the owl's neural circuits learn to make precise associations between auditory cues and locations in space. Photograph by Masakazu Konishi.

Foundation, and a multitude of private foundations, each focusing on neuroscience in a different way. How does a foundation make its mark among these diverse funders? Astute philanthropy avoids having its funds merely flow alongside the large stream of federal dollars. Rather, it develops a strategy to target its funds so they will be leveraged to have a disproportionate impact. This means meeting the specific needs of those excellent scientists one has worked so carefully to identify — in other words, filling the coverage gaps in the present funding environment that hold those scientists back.

In effect, the EFN has shouldered the added risk inherent in the out-of-the-box idea.

For the McKnight program, this has meant funding early exploratory studies for which there is not yet sufficient preliminary data to make the proposed project likely to meet the approval of an NIH review board. In effect, the EFN has shouldered the added risk inherent in the out-of-the-box idea. In this way, the McKnight program complements the more conservative funding priorities of the NIH, which are not intended to, but in effect end up favoring safe, staid science.

McKnight awardees say in near unison that they have used a McKnight Award to gather initial data for their most creative but fledgling project, which later garnered a much larger NIH grant. Furthermore, many awardees say this fledgling project in the years thereafter grew into the most productive line of research in their labs and may never have taken flight without McKnight support.

Precise numbers are difficult to come by, but Fischbach estimates the multiplier effect of McKnight dollars into future NIH funds at more than tenfold. A crucial ingredient in realizing this strategy has been to set the size of the award at just the right amount. The awards are small enough to allow a wide spread of the limited funds to many talented scientists, but they are large enough to start a new line of exploration and subsequently obtain federal funds.

3. RECOGNIZE TIMELY OPPORTUNITIES

The strength of the program is: it lets us change.

—Corey Goodman

While NIH is looking for interdisciplinary programs to support, there is minimal enthusiasm for funding transitions in the careers of established scientists. McKnight serves as one of the few resources for a basic scientist to develop the requisite skills and knowledge to investigate a clinical problem. This is one of the few means for attracting top basic scientists to the complexities of human diseases.

—Pat Levitt

Until the late 1990s, the McKnight program pursued the two-pronged approach that might be described as "Catch them while they're young" and "Catch them while they're changing." Young scientists who have finished their postdoctoral period and are striking out on their own are at their most creative but also the poorest when it comes to research funds. They yearn to break free of the science they trained with and want to make their mark by opening up a new area. Yet they lack a track record or preliminary data for these goals, so they cannot find the funding to start up fast. These scientists can apply for the McKnight Scholar Award, which has operated with few modifications since its creation in 1976.

More established scientists have a different funding problem. They see their creativity squeezed as the NIH's funding dynamic shoehorns them into continuing to do what they have always done. They can obtain funding for the established research they are known for, but when they try to venture into a new area for which they have no proven expertise, or even dare a career change into translational science, the NIH tends to deny funding.

These scientists in the past have applied for the Investigator and Senior awards. Starting in 1998 and 2000, respectively, the Technological Innovations Award and the Neuroscience of Brain Disorders Award have taken over this niche. Irrespective of the stage of a scientist's career, the new awards channel the drive for innovation toward technology development and brain disorders. Over the course of their careers, some basic neuroscientists develop ideas for devices or techniques that could overcome present hurdles in neuroscience research. Others develop an interest in translational science that grows out of a desire to apply their experience to a problem of human health. There are no funding sources for these ambitions. The new McKnight Awards tap this resource for ingenuity and motivation.

> *The ability to sense when an area is ripe for a push, combined with a boldness to seize the opportunity, is what distinguishes an exceptional scientist from a fine one.*

Why is it all-important to catch a scientist at just the right time? Scientists always can use money. The answer lies in the dynamics of scientific progress. While its overall process looks like an incremental one of building on prior knowledge, science does not actually evolve at a steady pace. It lurches forward in a leap when someone makes a pioneering discovery that brings the outlines of new territory within grasp. This draws in other people who fill in the new area with additional, competitive discoveries and move it forward rapidly.

Years later, a vein of exploration can slow down as scientists bump up against the limits of currently available technologies and the discoveries become increasingly smaller. At this stage, the field requires another leap to break through those limits and enable a fresh wave of inquiry, often at a more sophisticated level of analysis. The ability to sense when an area is ripe for a push, when a new technology makes a perennial question suddenly approachable, combined with the boldness to seize the opportunity, is what distinguishes an exceptional scientist from a fine one. And even an exceptional scientist makes a true breakthrough only a few times in a career.

Here are some examples of uncanny timing: The McKnight Scholar Award to Linda Buck came just as she had made her initial discovery of the existence of a thousand genes for olfactory receptors but was setting up her first independent laboratory at Harvard Medical School and needed unrestricted funds to analyze her data to make their impact clear. McKnight funded the early part of the work that earned her a Nobel Prize 13 years later.

Likewise, Roderick MacKinnon conducted his prize-winning work in part with McKnight funds. MacKinnon's second McKnight Award exemplifies a scientist's capacity for change. He was known for his ion receptor physiology, but in the mid-1990s he decided to veer away from that and tackle what no one had done before: solving the atomic structure of an ion channel by X-ray crystallography as a way of explaining at a molecular level how these protein pores manage to be selective for particular ions. He isolated and crystallized a potassium channel—a Herculean task for any membrane protein at the time—and in a brilliant series of papers analyzed how its structure explained the peculiarities of its function. In his Nobel Prize autobiography, MacKinnon wrote, "I have learned that most people do not like change, but I do. For me change is challenging, good for creativity, and it definitely keeps life interesting."

Many other well-timed funding opportunities pervade McKnight's record. Take Corey Goodman, who used McKnight money to switch from grasshopper anatomy to fruit fly genetics, or Larry Zipursky, who used McKnight money to leave behind bacterial genetics and spearhead the study of neuronal circuit development in the fly. Consider Tom Jessell, who used McKnight funds to change direction from neuropeptides to the study of transcription factors that define the specificity of motor circuits in mammals. Or consider Carla Shatz, who used McKnight funds to discover the spontaneous retinal waves in the fetal eye when the NIH proclaimed the project unlikely to work.

The Brain Disorders Award is too new to be able to look back on the long-term return of its investments, but its recipient roster is replete with scientists changing direction. Kimberly Huber used her background in basic synaptic plasticity to study the molecular basis of mental retardation. Paul Patterson used his experience in growth factor research and developmental biology to create an innovative mouse model for schizophrenia. It serves to test his idea that a pregnant mother's immune response to a viral infection might subtly damage the developing fetus's brain in such a way as to predispose it toward schizophrenia later in life. Likewise, Pat Levitt is using his basic background in cortical circuit formation to test new ideas in schizophrenia. Finally, consider Gary Yellen, a noted ion channel physiologist, who is trying to find out by which mechanism a ketogenic diet might help children with epilepsy.

More broadly, scientists attest to the EFN's skill in supporting them at the most productive times of their career. Many have said that the McKnight-funded work has been their best to date. Others even say the research would not have been done without this support, at least not in the same expedient and competitive way.

4. FOLLOW THROUGH

I am grateful to McKnight for its . . . consistent and continuing support of excellent neuroscience of many years.

—Arthur Karlin

*Many investigators with potential . . . have evolved into greatness in large
part because of the timely funding they received from McKnight.*
—Scott Small

Once a scientist has opened up a rich new vein of exploration, it makes sense to
keep the momentum by funding some of his or her best descendants and colleagues
so they can build the whole story. To continue a previous example, Dan Minor and
Jian Yang each received McKnight support to build on MacKinnon's precedent
and reveal atomic structures of additional ion channels, and they are succeeding.
The field of ion channel and receptor crystallography is still in its infancy. Many
formidable challenges of relevance to brain diseases still lie ahead, such as solving
the crystal structure of the massive gamma-secretase enzyme complex that pro-
duces the amyloid-beta peptide implicated in Alzheimer's disease.

More mature fields where McKnight has established a legacy of funding suc-
cessive generations of leaders include developmental neuroscience, with lineages
and colleagues of Goodman, Jessell, and Shatz; molecular mechanisms of
memory, with the progeny of Kandel; and systems neuroscience, going forward
from Evarts, Wurtz, Lisberger, and Squire. And even as Linda Buck and Richard
Axel opened up the molecular exploration of our sense of smell, McKnight has
supported this field by funding many of its most prolific contributors, such as
John Carlson, Catherine Dulac, Stuart Firestein, Jeffry Isaacson, Alex Kolodkin,
Gilles Laurent, Michael Lerner, Liqun Luo, Peter Mombaerts, John Ngai,
Gabriele Ronnett, and Leslie Vosshall.

The rationale for this sustained commitment to new lines of research goes
beyond a wish to stoke a small fire until it burns brightly. It is the awareness that
to understand how disease processes work, one needs first to understand the
normal process. This comes paired with an appreciation that animals with comparatively simple nervous systems offer a springboard to the more complex
human brain and that a surprising number of component parts and even basic
molecular mechanisms are conserved across the evolutionary spectrum.

5. ATTACH NO STRINGS

*The McKnight Endowment Fund gave me freedom to undertake risky
projects without writing fanciful justifications.*
—Masakazu Konishi

*I had the luxury to spend a couple of years exploring some brain areas in
the most unconstrained manner, looking for things I had no idea existed.*
—Gilles Laurent

After it has found superior scientists, the McKnight program exerts no control
over them and imposes no restrictions on them. The EFN trusts that the awardee

Leslie Vosshall (Scholar, 2001) studies the olfactory function of the fruit fly (lower left). Smell neurons running inside the antenna (green) are decorated with receptor proteins (paired red and blue subunits) that allow the fly to smell many different odors. Courtesy of Leslie Vosshall. Photograph of fly © Juergen Berger, Max-Planck Institut für Entwicklungsbiologie, Tübingen; 3-D animation by Joseph Alexander of The Rockefeller University.

will know best how to use the funds. It permits their use for whatever the awardee's research requires at the time, be it salaries, equipment, or supplies. In the spirit of investing in individuals, the foundation prohibits institutions from charging overhead costs on the award.

The EFN even permits awardees to "follow their nose"—that is, if in the course of studying their proposed topic they discover hints about another that are more intriguing, they may change direction. This is in contrast to federal funding agencies, which expect that funds be spent exactly as proposed in the application. Such restrictions can leave tantalizing new ideas to languish unfunded while a more conventional topic gets attention. More generally, the unrestricted nature of the award allows researchers to tackle difficult problems they might not otherwise.

The way McKnight Awards are administered reflects this philosophy as well. The EFN aims to set its scientists free to do what they do best—explore the brain—and to avoid distracting them with onerous administrative requirements.

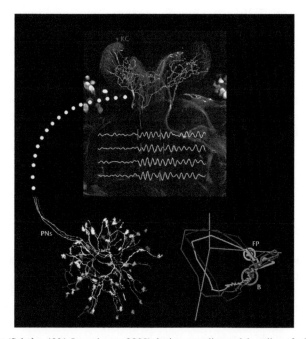

Gilles Laurent (Scholar, 1991; Investigator, 2000) depicts encoding and decoding of odors in a locust brain. Laurent studies the connectivity and spatio-temporal firing patterns of smell neurons, and traces the higher-order recognition of odor perceptions in a smell center called the mushroom body. Reprinted from *Neuron*, 48; Mazor and Laurent, "Transient Dynamics Versus Fixed Points in Odor Representations by Locust Antennal Lobe Projection Neurons," 661–673, © 2005, with permission from Elsevier. Image by Sarah Farivar, Glenn Turner, Ofer Mazor, and Gilles Laurent.

They need not file spending accounts or frequent progress reports. They are merely asked to describe the success, or failure, of their project at the end of their award and to present their data at the annual McKnight conference.

For their part, the EFN selection committees and boards also perform their duties unencumbered by bureaucracy. They do not handle the administration of the program through a separate EFN staff but leave it in the hands of professional staff at the McKnight parent foundation. This allows them to concentrate on awardee selection and programmatic issues.

6. Mix It Up

Many cross-field collaborations fade away very quickly — talk is cheap — but the McKnight Award allowed us to build an ongoing team that has already produced tangible results.

—A. David Redish

For any scientific field to expand and transcend its current limits, it is essential that it incorporate innovation from outside. Frequently, it is technological breakthroughs in quite separate areas that can lift up one's own field, once properly absorbed. This sounds self-evident in principle, but achieving interdisciplinary influx can be difficult in practice. Established funding mechanisms and the sociology of science can in effect promote an insular view that works against such reaching out. Resistance can be felt on the inside, by a neuroscientist who turns toward "newfangled" techniques, or from the outside, say by a geneticist who presumes to apply his reductionist approach to study human behavior. Supporting this interchange is a means by which The McKnight Foundation has become a motor for neuroscience innovation. Philanthropy can use its inherent flexibility to fulfill this role.

> *One of the strengths of the EFN has been its emphasis on engaging people who were not neuroscientists.*

One of the strengths of the EFN has been its emphasis on engaging people who were not neuroscientists. "Strictly speaking, some people were outsiders at the time they were funded, even though the committee felt they really should be neuroscientists," Fischbach recalled. "Neuroscience looked at them with skepticism. They were not appointed in neuroscience departments. At NIH, nobody on their respective study section knew them because they were in a new field. It took insight by the committees and courage by the foundation to go ahead with them. It was often they who established what is now the traditional way of conducting neuroscience."

One example is James Gusella, who received an Investigator Award in 1985 when he was an assistant professor in a genetics department. That was before genes for brain disorders had been identified but just after Gusella had found a marker for Huntington's, a clue that a gene for this devastating form of neurodegeneration could be pinpointed in a patient's DNA. The gene's eventual identification in 1993, by a consortium of scientists that included Gusella, created great excitement among neuroscientists. It became a moment of validation for the budding field of neurogenetics. It drove home their point that genetics could serve as an alternative route to understanding the mechanism of a disease.

This was a departure from the conventional approach, which tended to observe pathology in postmortem tissue of patients and work its way back from end-stage disease toward its original mechanism. By contrast, geneticists sidestep the phenotype (i.e., physical expression of a disease); they search for flaws in the DNA and then work forward from those flaws to unravel the molecular path to disease.

By now, neurogenetics has become an established subarea of neuroscience, and Gusella directs the Neurogenetics Unit at Massachusetts General Hospital. Other successes followed hard on the heels of the Huntington's gene—for example, the identification of genes for neurofibromatosis, for Lou Gehrig's disease, for Alzheimer's and related forms of dementia, for certain forms of mental retardation and epilepsy, for triplet repeat diseases, and, more recently, for Parkinson's disease.

Another example of scientists who influenced neuroscience from the outside can be found in Michael Greenberg, a molecular biologist who advanced the study of how neuronal activity leads to intracellular signal transduction and changes in gene expression. Or consider Stephen Lippard, a chemist who developed sensors that monitor the effect of neurotransmitters on metal chemistry inside the neuron. Lippard received a 2000 Technological Innovations Award. Other McKnight awardees have come from cell biology, physics, developmental biology, biochemistry, and engineering.

Of course, not only outside scientists bring new technologies to neuroscience. Enterprising neuroscientists do it, too, and have received McKnight support for those transitions. Examples include Fred (Rusty) Gage, a 1988 Investigator awardee, who in the late 1980s applied new methods of gene transfer to studies of the adult nervous system. This work pioneered studies of transplantation of genetically modified cells into adult brain tissue to try to repair damaged circuits and restore lost function. This area has become a focus in Parkinson's research and, to a lesser degree, in Alzheimer's. Likewise, Stephen Heinemann used McKnight funds in the early 1990s to establish in his neurobiology laboratory the new technology of creating genetically engineered mice, which has since become a mainstay of neuroscience.

7. THINK BROADLY

By funding many . . . investigators early in their careers and giving them the freedom to pursue their ideas, The McKnight Endowment Fund for Neuroscience has had a major impact on brain research in the United States, an impact far beyond the monetary value of the funds that have been disbursed.

—Robert Malenka

From the outset, the scientific founders of the McKnight neuroscience program and their counterparts at the foundation have agreed that they would approach their long-term goal of understanding diseases of memory in the broadest possible way. This contrasts with the strategy at some other private foundations started by families who are stricken with a particular disease. They tend to dispense their funds more narrowly to scientists who are expressly studying the disease at hand. The EFN's position is that important clues to diseases of the mind can spring up from the most unexpected corners of neuroscience, provided these seemingly remote areas advance rigorously to a mechanistic view of how their components interact among themselves and with other parts of the brain.

Important clues to diseases of the mind can spring up from the most unexpected corners of neuroscience.

In this spirit, the EFN has balanced its support between core areas of interest—mechanisms of neural development, synaptic function, plasticity—on the one hand and other areas of neuroscience on the other. As scientists learn more, these other areas that are seemingly remote from the brain and its diseases are proving to be connected at a mechanistic level after all.

An example for the program's breadth is 2001 McKnight Scholar Daniel Feldman. He has worked out how sensations of touch get processed in the brain. While studying this problem, Feldman has delivered a first-rate description of how the loss of sensation permanently tamps down the responsiveness of specific, identified synapses in the cortex of awake, behaving animals. In this way he has advanced the study of synaptic plasticity mechanisms. The relevance of this new work to brain disease is not clear yet, but it is likely that the fundamental mechanism Feldman has unraveled may prove relevant once scientists explore it further.

Consider some further examples of the program's wide reach. Emmanuel Mignot, who received a 2002 Brain Disorders Award, has studied narcolepsy as part of his interest in the molecular control of sleep. Narcolepsy is a bizarre and dangerous condition, in which people's sleep-wake cycle is in such disarray that they collapse into paralysis and REM sleep suddenly, especially in response to emotional excitement. Mignot had cloned the gene for narcolepsy in dogs and found that it codes for hypocretin, an excitatory neuropeptide involved in many functions, among them the control of food intake.

Intriguingly, people with certain variants of immune HLA alleles (the cell surface proteins that "present" antigens to patrolling immune cells) are particularly prone to narcolepsy. There is reason to suspect that narcolepsy is an autoimmune disease much like Type 1 diabetes, such that neurons that normally release hypocretin get attacked and die. (In childhood diabetes, insulin-producing cells in the pancreas die from autoimmune attack.) To find out the molecular context in which hypocretin functions, Mignot turned to zebrafish. This was the point of transition where Mignot needed McKnight support; coming from sleep medicine, he had no proven expertise in zebrafish biology or genetics.

This work is ongoing, but Mignot has already established that zebrafish sleep and that their sleep is controlled by neurochemicals that are conserved in humans. He has also located the hypocretin-releasing cells in the fish's brain. He first developed techniques for quantifying the animals' sleep rhythms and then made strains with disrupted sleep patterns, expecting that the mutant genes underlying the disrupted sleep will lead him toward additional players in the control of human sleep. It's not too farfetched to assume that research into peptides involved in eating and sleep control will eventually become relevant to aging. Old people, particularly people with Alzheimer's disease, are known to

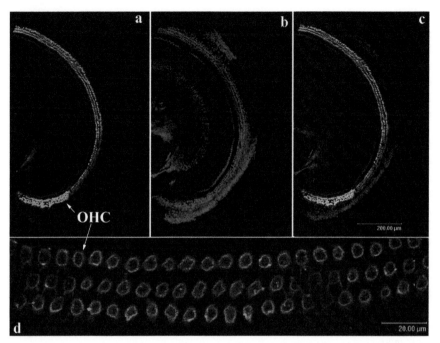

Peter Dallos (Senior Investigator, 1997) studies the molecular biology of hearing. Here, the organ of Corti in the inner ear of a gerbil is labeled for a motor protein on the hair cells of the cochlea (a, c, d, in green). Red are all cells of this organ, for comparison. Reprinted from the cover of *Mammalian Genome*, 14 (February 2003); in conjunction with an article by Zheng, Long, Matsuda, Madison, Ryan, and Dallos, "Genomic Characterization and Expression of Mouse Prestin, the Motor Protein of Outer Hair Cells," 87–96, © 2003, with kind permission of Springer Science and Business Media.

have disturbed sleep-wake cycles and get up in the middle of the night to eat or go shopping.

The neuroscience of hearing is another area that is receiving McKnight support. Long dominated by biomechanical and electrophysiological approaches, this field was late in transitioning to molecular biology. McKnight is nudging the process along. Examples include Peter Dallos (Senior Investigator, 1997), who identified prestin. This molecular motor protein sits on the outer hair cells of the ear's cochlea. It converts voltage changes into the force necessary for the hair cells to change their shape as they amplify a received auditory signal. And 2005 Brain Disorders awardees Stefan Heller and John Brigande are working to develop a stem-cell–based therapy for hereditary deafness.

Charles Zuker (Scholar, 1988) received McKnight funds for his highly regarded discoveries on taste receptors, as have a large contingent of the researchers who collectively have greatly advanced our understanding of the sense of smell. In addition, our sense of heat and conjoined perception of burning pain have become much better understood with fundamental discoveries by David Julius. A 1997 McKnight Investigator and now an EFN board member, Julius discovered that the receptor on sensory nerve fibers that reacts to the pungent ingredient of red chili peppers, capsaicin, is the same as the one that senses heat. Moreover, he found that the receptor for menthol also senses cold, illustrating why the peppers are aptly called "hot" and mints "cool."

This principle of scientific diversity has endured past the revisions made to the McKnight Awards program in the late 1990s. Beyond funding a wide range of topics, diversity has come to mean a balance between molecular, cellular, and systems approaches in the Scholar Award, and between imaging, recording, and sensing methods in the Technological Innovations Award. In the Brain Disorders Award, the selection committee strives for a balance between studies of neurologic and psychiatric disorders on the one hand, and mechanistic and diagnostic/treatment studies on the other.

8. For Returns: Expect the Unexpected

One sets out looking for A, but discovers B, which proves just as interesting, if not more.

—Fernando Nottebohm

The stewards of the McKnight neuroscience program know that dividends to knowledge from basic science will come in unpredictable, often surprising ways. Consequently, they invest in fundamental, enabling research. To be sure, basic science has already expanded our understanding of who we are as human beings and how our healthy brain functions. On the applied side, it is too early to claim that basic neuroscience has eased the human experience of brain disorders on a large scale. Viewed overall, translational research is just beginning to realize its potential for therapy development as it draws increasingly on the body of basic knowledge that has grown since the 1970s. A look back 20 years from now is more likely to be able to point to a range of new diagnostics and therapies, though cautious neuroscientists avoid the word "cure." That said, smaller examples of benefits of basic neuroscience are already springing up in most areas of brain disease. While The McKnight Endowment Fund has not funded these directly, its role in driving the broader transformation of neuroscience into a molecular discipline has helped bring them about.

> *Smaller examples of benefits of basic neuroscience are already springing up in most areas of brain disease. While The McKnight Endowment Fund has not funded these directly, its role in driving the broader transformation of neuroscience into a molecular discipline has helped bring them about.*

Experimental therapies in Alzheimer's disease, ranging from protease modulators to vaccination approaches, derive from molecular neuroscience and even more basic endeavors such as enzymology and immunology, respectively. Similar approaches for Parkinson's will follow suit as genes for this disease, which have been discovered only recently, are more thoroughly studied. Already, some of the underlying mechanisms, such as protein misfolding or disturbances in the cell's oxidative metabolism, are showing surprising overlap with the science of Alzheimer's, as do important aspects of amyotrophic lateral sclerosis (ALS) and triplet repeat diseases, such as Huntington's. Some experimental therapies for Parkinson's draw heavily on what is known about the action of growth factors, a field McKnight funded in its early days.

The treatment of brain seizures and epilepsy has become more precise with the discovery of the many different types of ion channel that exist in the brain. Increasingly, antiepileptic agents are having fewer side effects as they are targeted to more specific subclasses of receptors. Other drugs reduce seizures by promoting the efficacy of GABA, an inhibitory neurotransmitter. Similarly, schizophrenia treatment has seen an improvement with the introduction of new antipsychotic medications that target only one, not all, types of dopamine receptor. This could not have been done had basic scientists not discovered and described the different classes of dopamine receptors.

Even more surprisingly, perhaps, there are indications that certain antidepressants might act at the level of neurogenesis. Even in the early 2000s, few researchers would have thought of the trickle of new neurons that are generated in certain brain areas throughout life as influencing mental health. What functions adult neurogenesis might subserve, and how we might be able to manipulate it for therapy, has become an emerging focus in neuroscience research.

In the preclinical arena, an unexpected convergence is occurring between brain development and cancer. Scientists have realized that many of the molecules that are used in axon guidance are also used in the patterning of the vascular system when new blood vessels form and grow. Remarkably, signals that guide axons by repelling them from a certain path turn out to attract vascular cells. This becomes clinically relevant during the process of tumor angiogenesis, where blood vessels invade a small tumor to sustain its growth. In fact, these axonal guidance proteins form a parallel pathway in pathological blood vessel formation that may be as important as is that of VEGF, the card-carrying growth factor in blood vessel development, some neuroscientists now believe. Some of the newest cancer drugs act by choking off a tumor's blood supply, and they achieve a modest extension of patients' lives. Conceivably, this extension might become longer if additional drugs cut off signaling by the parallel set of axon guidance proteins as well.

Conversely, inhibitors of the axon guidance proteins might promote nerve regeneration from nerve injury by removing their repellent effect on regenerating axons. Many of these molecules were first discovered by McKnight awardees

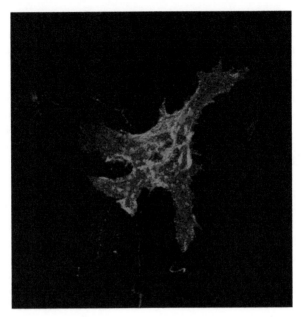

Tito Serafini (Scholar, 1996) showed that a single neuronal protein, neuroligin, can initiate a cascade of events that leads to the formation of functional synapses, even if the cell displaying neuroligin on its surface is not a neuron. Here, non-neuronal cells (green) are made to express neuroligin; neurons (red) grow over them and establish synaptic contacts. Courtesy of Tito Serafini.

Corey Goodman, Marc Tessier-Lavigne, and Tito Serafini. McKnight awardee Alex Kolodkin has participated in realizing their role in tumor angiogenesis. These scientists now are involved in biotech ventures in this area. On the flip side of this coin, the vascular growth factor VEGF shows promise in treating animal models of ALS.

Similarly, another protein of developmental neuroscience has proven important to cancer research. It is hedgehog, the protein that helps define the identity of neurons in the embryonic spinal cord. Tom Jessell and 1999 McKnight Scholar Henk Roelink have studied this protein. Hedgehog signals, too, are known to drive the progression of some types of cancer, and hedgehog antagonist drugs are now in clinical development.

While cancer drugs were not foremost on William McKnight's mind when he set aside funds for neuroscience, he certainly would have welcomed this unexpected clinical relevance to his philanthropy. And what his daughter, Virginia Binger, wanted the foundation to do—help find a cure for a disease—could come to pass via this unanticipated route.

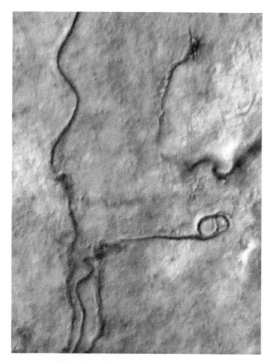

Using the fruit fly embryo, Alex Kolodkin (Scholar, 1995; Investigator, 2000) demonstrated that in the absence of crucial guidance proteins, developing motor axons in the periphery (brown) lose their way and extend aberrant projections toward muscles. Reprinted from *Cell*, 109; Terman, Mao, Pasterkamp, Yu, and Kolodkin, "MICALS, a Family of Conserved Flavoprotein Oxidoreductases, Function in Plexin-Mediated Axonal Repulsion," 887–900, © 2002, with permission from Elsevier.

9. CREATE A COMMUNITY OF PEERS

Pure and simple, [the McKnight conferences are] the best meetings I attend. I come back intellectually invigorated and with new ideas to test.

—Irwin Levitan

This is one of the few meetings that can successfully bring together the full spectrum of neuroscientists, ranging from neurodevelopment in worms to psychophysics in primates. The conference is a good opportunity to discuss science with colleagues and initiate new collaborations.

—Paul Slesinger

Now held at the Aspen Institute in the Colorado Rocky Mountains, the McKnight Conference on Neuroscience provides a forum that unifies the activities of the endowment fund. Each year, the EFN invites all current awardees and a changing selection of past awardees and special guests to share ideas in formal talks and plenty of informal discussion. The board plans each conference by balancing presentations by awardee whose grants are nearing the end of their run, with selected lectures by leading senior researchers.

> *The conference has built a community of peers in which the unifying element is a sense that all members are there because of their excellence, potential, and commitment to the field.*

Importantly, the board also selects a different brain disorder each year to receive special attention in a half-day workshop that begins with brief presentations by a leading clinician and a renowned researcher and then opens up to extensive discussion by the full audience. The workshop is unique in that it places clinical and research leaders jointly in front of a highly skilled audience for 2 hours of searching discussion. The scientists try to identify aspects of the disorder that might be amenable to study and new scientific insights and approaches that could be brought to bear on it. McKnight scientists report that the workshops stimulate them to think about how they can make their research more relevant to human health. Many carry new ideas back to their labs; some hatch plans for collaboration with fellow McKnight scientists.

Few other research philanthropies stage regular awardee conferences. It is more customary for a foundation to award a grant and request progress reports in return, but there is typically less of a focus on personal interaction among awardees. For the EFN, the conference has grown in importance over time. It has built a community of peers in which the unifying element is not age or topic area but a sense that all members are there because of their excellence, potential, and commitment to the field. In this way, scientists can reach across their own subject area for collaborations with any of the other McKnight fellows, knowing they will interact with a creative mind to match their own. Collaborations do indeed grow from discussions at the conference. For example, one by Michael Mauk (Scholar, 1989) and Steven Lisberger (Scholar, 1981) led to a paper in the journal *Science* about principles of motor learning in the cerebellum.

Moreover, the McKnight conference increasingly fills a new need that arose as the overall field of neuroscience burst at the seams. The annual conference of the Society for Neuroscience each year draws a growing number of people; it stood at more than 25,000 in 2006. At such large gatherings, it becomes nearly impossible for a researcher to keep up with the most important developments outside his or her own narrow area of study.

Offering quality rather than quantity, the McKnight conference fits the opposite profile. The deliberation going into selection of awardees combined with the planning of the conference program ensures that each talk is worth taking in because it represents the frontier in an area outside one's own. Often in such an environment, presentations about a seemingly distant subject area can prove unexpectedly relevant to a scientist's own research, and probing discussions ensue that plant the seeds for fresh projects.

The conference also devotes time for selected demonstrations of new technologies by recipients of the Technological Innovations Award and others. Taken together, the conference mirrors the general motto of the McKnight neuroscience program: small but consistently excellent.

The conference ambience is casual, with ample opportunity to interact socially and forge collegial friendships. Annual attendance over the course of one's award and then occasional visits thereafter help scientists deepen their connections with peers so they can make them last throughout a career. Young investigators, in particular, or those who did not attend elite institutions benefit from having ready access to leaders from the full spectrum of neuroscience.

"One cannot overestimate the value to a young scientist of realizing that one can really make a contribution and of the opportunity to establish connections with others sharing similar aspirations," said awardee Urs Rutishauser.

"The opportunity to attend this small conference is one of the most attractive features of this program for both committee members and awardees," Barondes said. "Spending a few days with about a hundred other excellent neuroscientists fosters warm personal interactions that facilitate new collaborations and ideas."

10. PRACTICE TRANSPARENCY, BALANCE, AND MERITOCRACY

[The McKnight program is] probably the most important source of funding in the field of neuroscience, apart from NIH and HHMI.
— Susan McConnell

Transparence, balance, and meritocracy are widely understood principles of good governance today. The EFN has realized them with an open selection process for all its awards, a geographic balance of directors, committee members, and awardees drawn from all over the country, and conflict-of-interest rules. Finally, term limits and rotation schedules for the selection committees, officers, and directors balance regular turnover with stability and help ensure objectivity.

SUMMARY

William L. McKnight was disheartened by the failing of his memory and that of his friends. He gave to other medical causes, too, but near the end of his life, helping people with memory loss was the goal he thought most worthy of his philanthropy. "A well-financed brain research organization to study the functions of the brain, the most important organ in the human body and the least known organ, could become a great gift to humanity," he wrote to his daughter and son-in-law.

In the years since his death, the problems of age-related memory loss and other brain disorders are not solved, but knowledge of the brain has advanced immeasurably, and William L. McKnight's contributions to brain science have proved to be an extraordinary gift.

IN THEIR OWN WORDS: TESTIMONIALS FROM AWARD RECIPIENTS

McKnight awardees responded to a survey in 2003 about the value of their award to their career. What follows are excerpts from some of the responses.

SARAH BOTTJER, UNIVERSITY OF SOUTHERN CALIFORNIA
SCHOLAR AWARD, 1985

I was a non-tenure-track assistant professor at UCLA at the time. My husband is a paleontologist and had a great job in Los Angeles, but I was having difficulty getting a "real" job (I was looking only in Los Angeles). Happily, the next year I was hired at the University of Southern California. Two things stand out: (1) the McKnight Award made me even more competitive and enhanced my ability to get a job in that situation, so it helped a young female scientist; (2) had I stayed at UCLA, the money would have been used to buy my intellectual freedom. I had started an entirely new field of investigation within the neurobiology of vocal learning in songbirds, and the McKnight Award helped increase the pace and quality of the work. Another plus was attending the meetings and getting to know many senior scientists and discussing my work with them and learning about new data, approaches, techniques, and so on.

STEVEN J. BURDEN, NEW YORK UNIVERSITY
SCHOLAR AWARD, 1980

I received my award when I was beginning my independent career as an assistant professor at Harvard Medical School. Funds from the McKnight Endowment Fund gave me an opportunity to obtain equipment, hire staff, and initiate projects that would otherwise have been delayed. We were able to analyze the topography of proteins in the postsynaptic membrane at neuromuscular synapses and to establish that Rapsyn, then termed 43 kd protein, was associated in equal stoichiometry with acetylcholine receptors (AchRs) and could be cross-linked to the AchR beta subunit. Fifteen years later, studies by others showed that Rapsyn regulates the spatial distribution of AchRs and is required to cluster AchRs in the postsynaptic membrane.

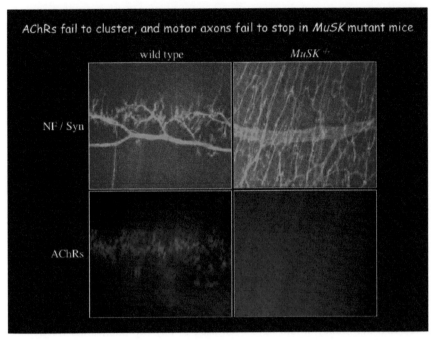

This image by Steven Burden (Scholar, 1980) shows how motor neurons require local clusters of transmitter receptors (red dots, bottom left) on muscle cells to be able to build synapses opposite those clusters (top left). Mice lacking a protein called muscle-specific receptor tyrosine kinase (MuSK) cannot form these specializations (red wash, bottom right), and motor axons grow over the muscle without making synapses. Courtesy of Steven Burden.

JOHN R. CARLSON, YALE UNIVERSITY
SCHOLAR AWARD, 1990;
INVESTIGATOR AWARD, 2000

I was extremely fortunate in receiving McKnight Awards at two critical stages in my career—a Scholar Award during a very early phase and an Investigator Award at a time when some exciting new possibilities had suddenly arisen in our laboratory by our isolation of the first insect odor and taste receptor genes. New funds were critically needed to capitalize on these opportunities. The funds were used to carry out some of the very best work of my career.

WILLIAM CATTERALL, UNIVERSITY OF WASHINGTON
SENIOR INVESTIGATOR AWARD, 1997; NEUROSCIENCE OF BRAIN DISORDERS AWARD, 2005

The support from The McKnight Endowment Fund was instrumental in helping to move my research in new directions. I had focused on the structure and function of sodium and calcium channels for many years. As the properties of these ion channels become increasingly well established, I have sought to redirect my research toward more integrated neuroscience objectives, first at the cellular and systems levels and eventually at the level of the whole animal and neurological disease. The McKnight support has helped me to make these transitions, studying the regulation of sodium and calcium channels in a more integrated context in intact neurons and in mouse genetic models of epilepsy. These studies have led to a clearer definition of the mechanism of regulation of these ion channels in neurons and to evidence for regulation of the input-output relationships of neurons by second messenger-activated protein kinase pathways that work through alteration of intrinsic ion channel inactivation processes. Altered regulation of these ion channels may contribute to epilepsy and other diseases of neuronal hyperexcitability.

E.J. CHICHILNISKY, SALK INSTITUTE FOR BIOLOGICAL STUDIES
SCHOLAR AWARD, 2000; TECHNOLOGICAL INNOVATIONS IN NEUROSCIENCE AWARD, 2004

In recent years, opportunities have arisen for collaborative ventures with outstanding colleagues to (1) develop multielectrode recording techniques for probing large-scale signaling in the retina, (2) examine how photoreceptor signals are conveyed through the retinal circuitry, and (3) develop computational models to understand visual processing in the retina. The McKnight Endowment Fund support made these collaborations possible, tremendously expanding the scope of my research and my own personal scientific horizons. In addition, the breadth and high quality of the annual McKnight conference exposed me to areas of neuroscience that I would not otherwise have appreciated, because I would not have known what work to pay attention to. The McKnight Endowment Fund has had a profound effect on my work.

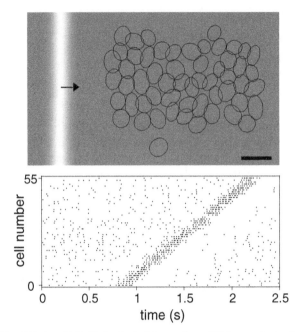

E.J. Chichilnisky (Scholar, 2000; Technological Innovations, 2004) records simultaneously from ensembles of individual neurons in the retina to understand how they cover a given visual field and encode their information. Reprinted from the *Journal of Neurophysiology*, 94; Frechette, Sher, Grivich, Petrusca, Litke, and Chichilnisky, "Fidelity of the Ensemble Code for Visual Motion in Primate Retina," 119–135, © 2005; used with permission.

HOLLIS CLINE, COLD SPRING HARBOR LABORATORY
SCHOLAR AWARD, 1991

The award came at a time in my career when I decided to make a significant change in the methodologies I used to study the structural plasticity of neurons in response to experience. . . . It allowed me to apply *in vivo* imaging methods to observe the structural changes in single neurons within the context of the intact developing brain. At the time, *in vivo* imaging studies permitted a rare view into the events of brain development and plasticity. We combined the imaging experiments with viral gene transfer methods to probe the molecular requirements for activity-dependent brain development. These studies opened up a new arena in developmental neuroscience that has led to a significant understanding of the mechanisms of axonal and dendritic arbor development and the formation of synaptic connections. Of particular note, we have observed that sensory experience enhances the structural development of axons, dendrites, and synapses. Furthermore, we have made significant progress in identifying genes and proteins required for normal brain development and circuit formation.

RONALD L. DAVIS, BAYLOR COLLEGE OF MEDICINE
SCHOLAR AWARD, 1984; INVESTIGATOR AWARD, 1988;
NEUROSCIENCE OF BRAIN DISORDERS AWARD, 2003

As I reflect on the advances made over my career so far, it is clear that support from The McKnight Endowment Fund underlies most of those that I consider the most important. The Scholar and Investigator awards provided critical support for my research as a junior neuroscientist. Now, the Neuroscience of Brain Disorders Award is providing the support necessary to extend my research beyond *Drosophila* and the mouse and specifically into issues of human cognition and disease. This is also a major new step for me. I hope these endeavors will be as successful as those supported by the Endowment Fund in the past and will yield insights into some of the complex diseases of the human mind.

VALINA DAWSON, JOHNS HOPKINS UNIVERSITY
NEUROSCIENCE OF BRAIN DISORDERS AWARD, 2004

The McKnight Award enabled me to pursue high-risk experiments with potential high impact. It enabled my lab to establish the utility of the functional cloning strategy we used to identify survival genes. We are discovering the cell signaling pathways that these novel proteins work through to protect neurons from injury. The McKnight Endowment Fund provides the seed funds to jump-start new creative science and allows investigators to rapidly generate discoveries and publications that can result in support from more traditional funding sources.

RICARDO DOLMETSCH, STANFORD UNIVERSITY
TECHNOLOGICAL INNOVATIONS IN NEUROSCIENCE AWARD, 1999; SCHOLAR AWARD, 2004

As a young postdoc in Michael Greenberg's lab at Harvard, I had an idea for a new technology that would allow us to look at the biochemical pathways active in neurons in much greater detail than ever before. The idea was to use a process called circular permutation to generate an enzyme that would be active only when a particular signaling pathway is active inside a neuron. This technology had important real-world applications; it had the potential to tell us about the timing of the biochemical changes that occur in neurons as they wire together and change their connections to produce functional circuits and memories. We needed seed money to embark on the project, and the Technological Innovations Award provided it. We seemed to miss a lot more than we hit, and we talked about abandoning the project, but we persisted. And then, one day, it worked. The cells containing our split enzymes miraculously turned blue, and we jumped up and down and went to see the Red Sox lose to the Yankees at Fenway Park. The technology is now being developed by several companies and promises to be a relatively inexpensive and highly sensitive method for screening for new drugs and for making diagnostic kits.

MARLA FELLER, UNIVERSITY OF CALIFORNIA AT SAN DIEGO
SCHOLAR AWARD, 2002

My McKnight Award enabled me to fund a postdoctoral researcher on a risky project and purchase a confocal that has allowed us to develop a novel imaging technique. I may have eventually pursued the same project, but it would have taken longer. By imaging the second messenger cAMP in neurons, we have been able to determine that cAMP levels both regulate and themselves change with depolarization. McKnight conferences have helped in several ways. First, I have heard some of the best talks I have ever heard in my life, which in general broadens my understanding of neuroscience but also teaches me effective communication strategies. Second, the intense interactions with such high-caliber colleagues is invaluable for generating new ideas for experiments as well as obtaining resources/reagents for my own work.

SCOTT FRASER, CALIFORNIA INSTITUTE OF TECHNOLOGY
SCHOLAR AWARD, 1984

The McKnight Award allowed me to complete the development of advanced imaging tools for following neuronal dynamics in intact living systems. Of my 16 *Science* and *Nature* papers, all but two owe their existence to the techniques I developed under McKnight. I can name two discoveries made with McKnight support: intravital imaging of growth cone dynamics in vertebrates and intravital imaging of cell lineages and migratory pathways.

FRED H. (RUSTY) GAGE, SALK INSTITUTE FOR
BIOLOGICAL STUDIES
INVESTIGATOR AWARD, 1988

I received the award for a project titled "Grafting of cells engineered to produce NGF to the brain." I had arrived from Sweden 3 years earlier and was initiating a new effort in my lab to apply recently developed methods in gene transfer and transplantation strategies to experimental questions in the adult nervous system. This award allowed me to focus the attention of a large portion of my lab on the new field of gene transfer to the CNS. We were able to (1) establish reliable and robust methods to infect autologous cells with biologically active genetic material, with a special focus on neurotrophic factors and neurotransmitter enzymes, and (2) demonstrate that transplantation of genetically modified cells to the adult and aged brain could repair damaged cells and circuits and restore impaired functions. This award encouraged me to persist in this new field despite significant skepticism. Gene transfer to the adult nervous system with genetically modified cells and viruses is now a standard tool in neuroscience.

PAUL GLIMCHER, NEW YORK UNIVERSITY
SCHOLAR AWARD, 1996; TECHNOLOGICAL INNOVATIONS IN NEUROSCIENCE AWARD, 1999

I was interested in a novel project for the study of the parietal cortex. I proposed to test the hypothesis that the posterior parietal cortex plays a critical role in decision making and that neurons in this area encode the desirabilities of actions. Major funders turned down this proposal, but my McKnight Scholar Award allowed me to pursue this hypothesis. The work became, without a doubt, my most important and innovative work. It is certainly the work I am most known for, and now, years later, it is the centerpiece of a growing subfield in neuroscience. My Technological Innovations Award allowed me to develop a completely novel technology for use in single-neuron recording. Before that award, all single-neuron studies in awake, behaving primates were conducted, quite literally, in the dark. It was common for a researcher to spend 2 or 3 months deciding where to place an electrode. We had begun to explore the possibility that ultrasonic imaging could be used to visualize electrodes and, quite literally, turn on the lights. We felt that given a few years and a couple of hundred thousand dollars, we could develop real-time ultrasonic electrophysiology — and we did! Now ultrasonic imaging is in growing use throughout the awake, behaving primate community. We expect that within the next decade, use of this technology will become universal — largely, if not entirely, due to the support of The McKnight Endowment Fund.

DWAYNE GODWIN, WAKE FOREST UNIVERSITY
TECHNOLOGICAL INNOVATIONS IN NEUROSCIENCE AWARD, 2001

My McKnight Award allowed me to build an infrastructure to pursue questions that were far outside of my funded research interests, particularly in the development of molecular approaches to questions that I have previously addressed only using physiological techniques. The combination of these has allowed me to achieve a new level of vertically integrated inquiry into basic mechanisms of ion channel physiology.

ULRIKE HEBERLEIN, UNIVERSITY OF CALIFORNIA AT SAN FRANCISCO
INVESTIGATOR AWARD, 2000

My laboratory studies how drugs of abuse act in the nervous system to change behavior. We approach this complex problem in a simple model organism, the fruit fly *Drosophila melanogaster*. Funding from McKnight allowed me to pursue this unconventional approach, which was originally received with substantial skepticism by traditional funding agencies. The award funded a project aimed at mapping regions in the fly's brain that regulate ethanol-induced behaviors. This unbiased approach led to the discovery that insulin-producing cells in the fly's brain regulate ethanol intoxication. There is growing evidence that insulin plays important roles in the mammalian nervous system. The observation that insulin may have a direct effect on reward systems in the context of food reward suggests the exciting possibility that it also regulates drug reward and reinforcement. Many drugs that target the insulin pathway exist, opening the door for novel treatment strategies for alcoholism.

CHRISTINE HOLT, UNIVERSITY OF CAMBRIDGE
SCHOLAR AWARD, 1986

The award had a huge impact on my career as a scientist. It supported my early work characterizing the development of retinal ganglion cells in the visual system. My collaborator and husband, Bill Harris, gave me some space in his laboratory, but I had a soft-money position that provided neither space nor setup funds. It was quite hard to get established as a woman scientist at that time, and in my position it was particularly difficult to become independent. The McKnight Award allowed me to hire a postdoctoral fellow and helped me begin to establish my own group and independent research program. As a prestigious award, it gave weight and significance to the work I was doing and led to further support and eventually to a faculty position. I believe my study of retinal ganglion cells as they send out their axons and navigate to their targets in the brain was the first such study in a verte-brate, and it provided the foundation of many of our studies thereafter. Without the McKnight Award, I'm not sure I would have stayed in science. It gave me con-fidence to continue the struggle to balance career and family.

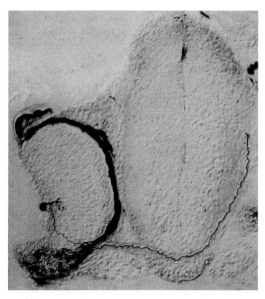

This photomicrograph from the lab of Christine Holt (Scholar, 1986) shows the visual pathway of a frog embryo 36 hours after fertilization. Filled with dye from a microelectrode, the black retinal ganglion neuron displays its complete morphology, with a round cell body in the eye (left), small dendrites nearby, and a long axon that traces its way unerringly through the optic nerve and deep into the midbrain. Courtesy of Christine Holt. A black-and-white version of this image appeared in the *Journal of Neuroscience*, 9; Holt, "A Single-cell Analysis of Early Retinal Ganglion Cell Differentiation in *Xenopus*: from Soma to Axon Tip," 3123–3145, © 1989 by the Society for Neuroscience; used with permission.

ARTHUR KARLIN, COLUMBIA UNIVERSITY
DISCRETIONARY AWARD, 1986; SENIOR INVESTIGATOR AWARD, 1994

My first McKnight Award permitted me to purchase the electrophysiological equipment we needed to match our teaching potential at our summer course in neurobiology at the Marine Biological Laboratory, which I ran from 1985 to 1989. Among the 50 students who used this equipment under the tutelage of outstanding instructors, many have become effective apostles of neurobiology. The 1994 award came at a time when I was developing our substituted-cysteine-accessibility method. In particular, it crucially supported salary and equipment for a postdoctoral scientist, Gary Wilson, who was working on the variation of the method that permitted the location of channel gates. Our approach has been widely adopted in the study of receptors, channels, and transporters. Once again, McKnight support at a critical time had a regenerating effect over a wide range of research.

RICH KRAUZLIS, SALK INSTITUTE FOR BIOLOGICAL STUDIES
SCHOLAR AWARD, 2000; TECHNOLOGICAL INNOVATIONS AWARD, 2006

I used the McKnight Scholor Award to (1) purchase additional single-unit recording equipment, (2) purchase materials and develop a new system for presenting visual stimuli for our experiments, and (3) provide support for postdocs in the lab. The new visual display system is a crucial component of our lab, and we wouldn't be able to design and perform the experiments we are doing now without it. Two of our most important discoveries were done with McKnight support: (1) Showing that neurons in the superior colliculus, a structure traditionally known for its role in the control of saccades, could also predict the choices made by pursuit eye movements. This showed that the two motor systems might use a common neural mechanism for selecting targets. (2) Showing that the superior colliculus plays a causal role in target selection for both pursuit and saccadic eye movements by applying weak microstimulation as the choice is being made. The surprising thing is that the effect on pursuit is based on where the target is located, not on the direction of the movement required to follow the target, thus identifying an effect on target selection itself separate from any effect on the motor plan.

IRWIN LEVITAN, UNIVERSITY OF PENNSYLVANIA SCHOOL OF MEDICINE
SENIOR INVESTIGATOR AWARD, 1997; NEUROSCIENCE OF BRAIN DISORDERS AWARD, 2004

The first award allowed me to take a flyer, exploring the modulation of an ion channel in which I had long been interested but had not had the opportunity to work with previously. The results of this study were very interesting and surprising, and, in fact, formed the basis for my second McKnight Award. I have become very interested in translational research but would not have been able to undertake my current efforts in that area without my Brain Disorders Award.

PAT LEVITT, VANDERBILT UNIVERSITY
NEUROSCIENCE OF BRAIN DISORDERS AWARD, 2002

I have said that the McKnight Award has been the principal supporter of my midlife crisis. It provided the financial support to introduce new technologies into my laboratory and to initiate relatively risky experiments to test a new biologically based hypothesis of schizophrenia. The McKnight Award is one of the few resources for a basic scientist to develop the requisite skills and knowledge to investigate a clinical neuroscience problem. This is one of the few means for attracting top basic scientists to address the complexities of human diseases.

MICHAEL MAUK, UNIVERSITY OF TEXAS, HOUSTON, MEDICAL SCHOOL
SCHOLAR AWARD, 1989

McKnight support allowed me to do two key things. First, I was able to complete my contributions to a project demonstrating the necessity of calmodulin and protein CaM kinase II for long-term potentiation. This work was published in *Nature* in 1992 and has been cited almost 600 times since. Second, I was able to keep my lab running at a time when federal funding was difficult to come by. I would have been unable to work without the McKnight Award. It literally saved my career in science. The conferences were a virtual who's who in neuroscience. One of the most important collaborative projects I have accomplished (with Steve Lisberger) was born out of discussions at the meetings.

EMMANUEL MIGNOT, STANFORD UNIVERSITY
NEUROSCIENCE OF BRAIN DISORDERS AWARD, 2002

I had just cloned the canine narcolepsy gene and found the cause of human narcolepsy. Then I needed to revamp my laboratory to start completely new projects. I became convinced that one of the ways I could move to the next step was to start a research program in zebrafish. Thanks to the McKnight Award, this project was started and is now in full swing. We isolated and characterized the hypocretin and histamine systems in the zebrafish with this award. It is now being used to find new genes affecting the pathway. The speakers at the McKnight conference convinced me further that functional genomics and animal models were the way to go for my long-term goals — finding why and how we sleep.

PARTHA MITRA, COLD SPRING HARBOR LABORATORY
TECHNOLOGICAL INNOVATIONS IN NEUROSCIENCE AWARD, 2000

I received my award jointly with Richard Andersen at Caltech. I continued the award project after being hired at Bell Laboratories. The award was instrumental in facilitating my career change from a theoretical physicist to a theoretical and experimental neuroscientist. It provided critical bridging funds and also allowed me to have a substantial collaboration with Richard doing primate electrophysiology. Concretely, the award allowed me to hire a postdoc, purchase equipment, and travel between Caltech and Bell Laboratories. In the course of the work pursued with McKnight support, we made and further established a salient discovery that may prove to be critical in the future development of neural prosthetic devices. We found that local field potential measurements from electrodes implanted in the brain show behavioral tuning and may be used to extract information to drive a prosthetic device.

FERNANDO NOTTEBOHM, ROCKEFELLER UNIVERSITY
SENIOR INVESTIGATOR AWARD, 1997

My laboratory focuses on the basic biology of vocal learning and neuronal replacement, using birds as a model. I used McKnight support to bring into the laboratory a molecular biologist who would look for changes in gene expression associated with changes in a bird's predisposition to learn. In retrospect, I jumped the gun. Our approach did not yield any breakthroughs. But that work kept alive in the lab the interest in molecular solutions, which now is yielding some very nice results. The molecular biologist I hired, Thierry Lints, was drawn into the behavioral work that was then flourishing in the laboratory and contributed significantly to ongoing studies that led us to look at vocal learning in a new manner that relies heavily on 24-hours-a-day sound recordings and computer analysis of the data. These tools really had to be in place before we could make a strong case for causal links between vocal learning and changes in gene expression. Dr. Lints also suggested the use of laser capture microscopy, which in conjunction with microarrays produced in my laboratory made us aware of the importance of a key gene, UCHL1, in the biology of replaceable neurons. Our present interest in the role UCHL1 might have in neuronal replacement can be seen as the delayed result of the 3 years of support from McKnight.

BARBARA RANSCHT, BURNHAM INSTITUTE
SCHOLAR AWARD, 1989

One research technician and I were working away on defining a new gene putatively involved in axon guidance. The receipt of the prestigious McKnight Award represented an endorsement of our research and spurred my enthusiasm and effort in the lab. We soon published our findings on the structure and expression of a novel (and still unique) GPI-linked cadherin molecule. We now know that this truncated cadherin regulates the patterning of axon projections extended by motor neurons and retinal ganglion cells. The award not only contributed to understanding molecular mechanisms of axon guidance but established interactions with scientists that directly influenced my thinking and future research.

A. DAVID REDISH, UNIVERSITY OF MINNESOTA
TECHNOLOGICAL INNOVATIONS IN
NEUROSCIENCE AWARD, 2002

The award allowed me to create an engineering collaboration team with my co-investigators, Art Erdman and Babak Ziaie, which is still ongoing and has (I believe) a great future. The award was for the development of new, miniaturized recording technology that would allow us eventually to record neural ensembles wirelessly. We've already developed a new amplifier capable of performing DC-noise rejection and the requisite amplification in a small, integratable area. We also built, implanted, and tested a motorized four-electrode microdrive. In a sense, the major contribution of the award was to allow us to create this collaborative team, to show that we can work together, and to make the cross-disciplinary connection. Many cross-field collaborations fade quickly (talk is cheap), but the McKnight Award allowed us to build an ongoing team that has already produced tangible results.

LORNA ROLE, COLUMBIA UNIVERSITY
SCHOLAR AWARD, 1988; INVESTIGATOR AWARD, 1994

Early in one's career, the energy and ideas are abundant and one is anxious to pursue new approaches. The timing of the Scholar Award enabled me to take some risks in my work that I would not have been able to pursue without inde-pendent support. I was able to pursue new molecular deletion approaches. Our reports using subunit-targeted antisense oligonucleotide-mediated deletion was one of the first to probe ion channel subunit composition with this more "in situ" technique. The Scholar Award enabled me to stretch into trying these novel techniques and supported my laboratory in the purchase of equipment, supplies, and personnel without whom the work would not have been done. The Investiga-tor Award also enabled an independent foray into a novel course of studies — the development and modulation of central cholinergic synapses. We contributed to changing the concept of CNS nAChRs to that of important and powerful presyn-aptic regulators of gain control.

URS RUTISHAUSER, MEMORIAL SLOAN-KETTERING CANCER CENTER
SCHOLAR AWARD, 1977

When I became a McKnight Scholar, the first cell adhesion molecule had just been identified, and it became possible to design experiments to study the role of this adhesion in development. This receptor was called a neural cell adhesion molecule (NCAM), and the logical substrate for studying it was the nervous system. My background at the time was primarily protein biochemistry and cell biology. The major impact of the fellowship on my career was the intellectual input from the yearly meeting of the Scholars and senior scientists. In making the transition to studying NCAM (and subsequently other aspects of cell interac-tion) on neurons, it was invaluable for me to hear and interact with researchers trained as neuroscientists. This input allowed me to gauge the status of a very broad and complex field. And of course, one cannot overestimate the value to a young scientist of realizing that one can really make a contribution and of the opportunity to establish connections with others sharing similar aspirations. The dividends of my fellowship have also paid off in recent years as I made another transition from developmental neuroscience into the role of plasticity in the mature vertebrate CNS, including physiology and behavior. It was very useful to draw on my initial exposure to the full range of neuroscience disciplines covered by the McKnight program.

GARY RUVKUN, HARVARD MEDICAL SCHOOL
INVESTIGATOR AWARD, 1991

The McKnight Award occurred at the nadir of federal funding, when all I was hearing was rather negative reviews of my work on *C. elegans* neurogenesis. Many of my peer group were driven out of science. Like an oasis, the letter from McKnight arrived with at least some hope of patronage from an elite scientific advisory board. This was a huge shot in the arm to my self-confidence, not to mention very crucial funding to my lab. The work funded by McKnight was rather high risk. We activated the neural specification gene UNC-86 by fusing it to VP16 and showed that this hyperactivated downstream genes. The McKnight Endowment Fund supplied four key resources: (1) funds for a risky experiment that paid off, (2) funds during a periodic federal funding downturn, (3) the cachet of the scientific advisory board endorsing the work in my lab, and (4) a real boost in my confidence.

CLIFFORD SAPER, HARVARD MEDICAL SCHOOL
SCHOLAR AWARD, 1983

The McKnight Award allowed me to follow a major interest that was more innovative than what I was working on at the time: understanding how the wake-sleep system controls the cerebral cortex. It gave me the flexibility to do some anatomical studies I was not otherwise funded to do and to begin some physiological studies that I did not have the equipment to pursue without the McKnight funding. These interests have dominated the last 10 years of my career. The McKnight support led to the first accurate map of the cholinergic system in the rat brain, the first systematic map of cortical afferents from the basal forebrain and hypothalamus in rat brain, and a highly cited review on diffuse cortical afferent systems in the *Handbook of Physiology*. I think the most important work that has come from my lab is identification of specific circuitry in the brain that regulates sleep and wakefulness, and I trace this back to the McKnight Award, which provided the preliminary results.

This in-situ hybridization autoradiograph from Clifford Saper (Scholar, 1983) shows the distribution of receptors for orexin-2 in a rat brain. This neurotransmitter controls wake-sleep cycles, appetite, motivation, and is involved in the sleep disorder narcolepsy. Reprinted from the *Journal of Comparative Neurology*, 435; Marcus, Aschkenasi, Lee, Chemelli, Saper, Yanagisawa, and Elmquist, "Differential Expression of Orexin Receptors 1 and 2 in the Rat Brain," 6–25, © 2001, with permission from Wiley-Liss, Inc., a Wiley Company.

MARC TESSIER-LAVIGNE, GENENTECH
SCHOLAR AWARD, 1991; INVESTIGATOR AWARD, 1994

The Scholar Award gave me the financial freedom to take a big risk, namely to try to identify through biochemical purification an axonal chemoattractant in the spinal cord. At the time, no axonal chemoattractants were known, and I was determined to try to identify one that we had been studying. But this was too risky a project for federal support, so I turned to The McKnight Endowment Fund. Receiving the Scholar Award meant I could forge ahead with the purification of the factors we ended up calling Netrins, work that helped advance our understanding of axon guidance mechanisms in a significant way. Without McKnight support, it would have been literally impossible to take this approach.

RICHARD F. THOMPSON, UNIVERSITY OF SOUTHERN CALIFORNIA
SENIOR INVESTIGATOR AWARD, 1985, 1988

I was engaged in very serious disputes with certain cerebellar physiologists about the role of the cerebellum in learning and memory. I was able to develop new approaches immediately, such as use of reversible inactivation to localize sites of memory storage. Among the work for which my lab is best known is a discovery made with McKnight support—localization of a memory trace for classical conditioning of discrete responses in the cerebellum, showing that at least some types of memory traces can be localized. McKnight's impact has been much greater than the actual amount of funding it provides because of the flexibility and innovative nature of the program.

GARY WESTBROOK, OREGON HEALTH & SCIENCE UNIVERSITY
INVESTIGATOR AWARD, 1988

I think what has made The McKnight Endowment Fund so important to neuroscience, and influential well beyond its resources, is the support of investigators at an early stage in their careers. This strategy encourages talented individuals to follow their ideas, and it ensures loyalty in former recipients. I credit McKnight with an important role in our work on the dynamics of glutamate action at central synapses, work published in *Nature* in 1990 and 1992. We developed methods for rapid application of ligands to patches of nerve cell membranes that allowed us to apply quantitative kinetic methods to ligand-gated channels. These techniques seem rather trivial now, but they really opened up biophysical studies of ligand-gated channels.

MONTE WESTERFIELD, UNIVERSITY OF OREGON
INVESTIGATOR AWARD, 1991

The award made a huge difference to my research and maybe to the field in general. At that time it was nearly impossible to obtain federal funding for mutant screens in zebrafish. The McKnight Award gave me flexibility to try something off the beaten path. Our screens worked, we identified interesting mutants and subsequently cloned the affected genes, and now zebrafish mutant screens are widely accepted as a useful and productive method for gene discovery and functional analysis.

ANNE B. YOUNG, MASSACHUSETTS GENERAL HOSPITAL
INVESTIGATOR AWARD, 1982

I received the award when I was interested in applying newly developed autoradiographic techniques to the study of neurotransmitter receptors in the human brain. Without McKnight support, our initial work would have been very much delayed. Our subsequent work on the postmortem human brain and animal brain allowed us to develop a model of basal ganglia function that is still widely accepted.

Historical Highlights

1975	William L. McKnight asks The McKnight Foundation to help evaluate Dr. Edwin Boyle's brain and memory research at the Miami Heart Institute, to which McKnight has given $2.5 million over 5 years. A team of consultants, headed by Fred Plum, recommends against supporting the research. Russell V. Ewald, the foundation's executive vice president; Virginia Binger, its president; and her husband, James H. Binger, want to establish a neuroscience program in honor of William L. McKnight, so the foundation invites Plum to submit a proposal.
1976	The McKnight Foundation approves $750,000 a year for Plum's proposal, which includes a three-part program of awards, a conference, and a McKnight Laboratory for Brain Science and Memory Research. Julius Axelrod agrees to chair a review committee to shape the program further. The committee includes Samuel Barondes, Edward Evarts, Seymour Kety, James McGaugh, and Plum. The awards and conference are established, but the laboratory is not.
1977	The foundation gives its first awards: nine Research Awards (later named Senior Investigator Awards) and five Scholar Awards. The first review committees include:

> *Scholars:* Plum, Barondes, Nelson Goldberg, McGaugh, Snyder
>
> *Research Awards:* Axelrod, Evarts, Murray Jarvik, Kety, Oliver Lowry, Eugene Roberts

1978	William L. McKnight dies on March 4.

117

1980   First McKnight Conference on Neuroscience is held at Spring Hill Conference Center in Minnesota. The conference is held every other year in various locations for the first two decades.

1981   The foundation creates the Neuroscience Development Awards (later Investigator Awards) for midcareer scientists.

1986   The McKnight Endowment Fund for Neuroscience (EFN) is founded. The first directors are: Plum (president), Axelrod, Barondes, Goodman, Ann Graybiel, Kandel, Kety, and Charles Stevens, with Russell Ewald and Cynthia Boynton representing the foundation.

1987   Barondes succeeds Plum as chair of the Scholar Awards Committee, and Kandel succeeds Axelrod as chair of the Senior/Investigator Committee. Cynthia Boynton succeeds her mother, Virginia Binger, as president of The McKnight Foundation. Boynton remains on the EFN board.

1989   Plum, Axelrod, and Kety leave the board, and Barondes succeeds Plum as president. Ewald retires from The McKnight Foundation, and his successor, Michael O'Keefe, replaces him on the board.

1990   Goodman succeeds Barondes as chair of the Scholar Awards Committee.

1991   Torsten Wiesel is invited to join the EFN board.

1993–1994   Abt Associates conducts a professional evaluation of the neuroscience program, praising the quality of the science supported but recommending a few administrative changes, such as term limits for board members.

1994   Term limits are instituted for board members, effective 1996. The McKnight Foundation approves $10 million to the EFN over four years, 1995–1998. Research and Development awards are renamed Senior Investigator and Investigator awards.

1995   Gerald Fischbach is elected to the board.

1996   Stevens and Graybiel rotate off the board. Fischbach chairs the Senior/Investigator Awards Committee.

1997   Thomas Jessell is elected to the board. The foundation extends its commitment of $2.5 million per year through 2001.

1998   Kandel rotates off the EFN board. Carla Shatz joins the board. Jessell succeeds Goodman as chair of the Scholar Awards Committee. Torsten Wiesel is elected president beginning in 1999. Goodman, Jessell, and Shatz are named to a planning committee for the future of the EFN. The proposal calls for a new Technological Innovations in Neuroscience Award, replacing the Senior Investigator Award and beginning in 1999; a Memory and Brain Disorders Award, replacing the Investigator Award and beginning in 2001; and an annual conference that includes a workshop on brain diseases and disorders. The foundation approves a 10-year grant to the EFN, beginning in 2001. Lubert Stryer agrees to join the board and chair the Technological Innovations Awards committee.

1999   Rip Rapson succeeds Michael O'Keefe at the foundation and joins the EFN board. Noa Staryk succeeds her mother, Cynthia Boynton, as McKnight Foundation president. Boynton remains on the EFN board. Larry Squire is elected to the board and agrees to chair the committee for the new Memory and Brain Disorders (later Neuroscience of Brain Disorders) Awards.

2000   McKnight Conference on Neuroscience becomes annual and is held in Aspen, Colorado. Fischbach leaves the board. Senior/ Investigator Committee disbands after selecting a final group of investigators for 2000–2002.

2001   Goodman becomes EFN president, with Shatz as vice president.

2002   Eric Nestler joins the board.

2003   Huda Zoghbi joins the board.

2004   David Julius joins the board. Erika Binger succeeds her cousin, Noa Staryk, as McKnight Foundation chair.

2005   Peggy J. Birk becomes interim president of The McKnight Foundation, succeeding Rapson. Pat Binger replaces Cynthia Boynton as McKnight Foundation representative to the EFN. Shatz becomes EFN president, and Goodman becomes vice president. David Tank joins the board and succeeds Stryer as chair of the Technological Innovations Committee.

2006   The EFN celebrates its 20th anniversary. The McKnight Foundation names Kathryn Wolford as its new president, and Birk returns to the parent foundation board.

Appendix I
About the Awards

For more information, see www.mcknight.org/neuroscience.

McKNIGHT SCHOLAR AWARDS

The McKnight Scholar Awards encourage neuroscientists in the early stages of their careers to focus on disorders of learning and memory. These awards support young scientists who hold the M.D. and/or Ph.D. degree, who have completed formal postdoctoral training, and who demonstrate a commitment to neuroscience. The endowment fund especially seeks applicants working on problems that, if solved at the basic level, would have immediate and significant impact on clinically relevant issues. The Scholar Awards have been given annually since 1977. Each year up to six Scholars are selected to receive support for 3 years at $75,000 per year.

2006 SCHOLAR AWARDS SELECTION COMMITTEE

Thomas M. Jessell, Chair
Thomas Albright
Cori Bargmann
Pietro De Camilli
Allison J. Doupe
Craig Jahr

McKNIGHT TECHNOLOGICAL INNOVATIONS IN NEUROSCIENCE AWARDS

Established in 1999, these awards support scientists working on new and unusual approaches to understanding brain function. The program seeks to advance and enlarge the range of technologies available to the neurosciences and does not support research based primarily on existing techniques. The fund is especially interested in how technology may be used or adapted to monitor, manipulate, analyze, or model brain function at any level, from the molecular to the entire organism. Up to four awards are given each year, and each award provides $100,000 annually for 2 years.

2006 McKnight Technological Innovations in Neuroscience Awards Selection Committee

David Tank, Chair
Laurence Abbott
Catherine G. Dulac
Stephen Heinemann
David Julius
Michael Shadlen

McKNIGHT NEUROSCIENCE OF BRAIN DISORDERS AWARDS

These awards were established in 2000 to support neuroscientists who are working to apply recent discoveries about the brain and nervous system to solving the problems of neurological and psychiatric diseases. They are designed to stimulate innovative approaches that might lead to therapies and cures. Collaborative projects between basic and clinical neuroscientists are welcomed, as are proposals that help link basic with clinical neuroscience. Up to six awards are given each year, each providing $100,000 annually for 3 years.

2006 McKnight Neuroscience of Brain Disorders Awards Selection Committee

Larry Squire, Chair
David J. Anderson
Charles Gilbert
Jeremy Nathans
Eric Nestler
Chris Walsh
Huda Zoghbi

PAST AWARDS

McKnight Senior Investigator Awards, for highly experienced scientists, were discontinued in 1997. McKnight Investigator Awards, for midcareer scientists, were discontinued in 2000. These two awards were replaced with the Technological Innovations and Brain Disorders awards.

Appendix II
McKnight Awardees, 1977–2006

McKNIGHT SCHOLAR AWARDS

2006–2008

Thomas Clandinin, Ph.D., Stanford University Medical School
James DiCarlo, M.D., Ph.D., Massachusetts Institute of Technology
Florian Engert, Ph.D., Harvard University
Youxing Jiang, Ph.D., University of Texas, Southwestern Medical Center
Tirin Moore, Ph.D., Stanford University Medical School
Hongjun Song, Ph.D., Johns Hopkins University School of Medicine
Elke Stein, Ph.D., Yale University

2005–2007

Matteo Carandini, Ph.D., Smith-Kettlewell Eye Research Institute
Miriam Goodman, Ph.D., Stanford University
Bernardo Sabatini, M.D., Ph.D., Harvard Medical School
Aravinthan Samuel, Ph.D., Harvard University
Nirao Shah, M.D., Ph.D., University of California, San Francisco
Athanossios Siapas, Ph.D., California Institute of Technology

2004–2006

Ricardo Dolmetsch, Ph.D., Stanford University
Loren Frank, Ph.D., University of California, San Francisco
Rachelle Gaudet, Ph.D., Harvard University
Z. Josh Huang, Ph.D., Cold Spring Harbor Laboratory

Kang Shen, M.D., Ph.D., Stanford University
David Zenisek, Ph.D., Yale University

2003–2005

Michael Brainard, Ph.D., University of California, San Francisco
Joshua Gold, Ph.D., University of Pennsylvania School of Medicine
Jacqueline Gottlieb, Ph.D. Columbia University
Zhigang He, Ph.D., Children's Hospital
Kristin Scott, Ph.D., University of California, Berkeley

2002–2004

Aaron DiAntonio, M.D., Ph.D., Washington University
Marla Feller, Ph.D., University of California, San Diego
Bharathi Jagadeesh, Ph.D., University of Washington
Bingwei Lu, Ph.D., The Rockefeller University
Philip Sabes, Ph.D., University of California, San Francisco
W. Martin Usrey, Ph.D., University of California, Davis

2001–2003

Daniel Feldman, Ph.D., University of California, San Diego
Kelsey Martin, M.D., Ph.D., University of California, Los Angeles
Daniel Minor, Jr., Ph.D., University of California, San Francisco
John Reynolds, Ph.D., Salk Institute for Biological Studies
Leslie Vosshall, Ph.D., The Rockefeller University
Anthony Wagner, Ph.D., Massachusetts Institute of Technology

2000–2002

John Assad, Ph.D., Harvard Medical School
Eduardo Chichilnisky, Ph.D., Salk Institute for Biological Studies
Frank Gertler, Ph.D., Massachusetts Institute of Technology
Jeffry Isaacson, Ph.D., University of California, San Diego
Richard Krauzlis, Ph.D., Salk Institute for Biological Studies
H. Sebastian Seung, Ph.D., Massachusetts Institute of Technology
Jian Yang, Ph.D., Columbia University

1999–2001

Michael Ehlers, M.D., Ph.D., Duke University Medical Center
Jennifer Raymond, Ph.D., Stanford University School of Medicine
Fred Rieke, Ph.D., University of Washington
Henk Roelink, Ph.D., University of Washington
Alexander Schier, Ph.D., New York University School of Medicine
Paul Slesinger, Ph.D., Salk Institute for Biological Studies
Michael Weliky, Ph.D., University of Rochester

1998–2000

Paul Garrity, Ph.D., Massachusetts Institute of Technology
Jennifer Groh, Ph.D., Dartmouth College
Phyllis Hanson, M.D., Ph.D., Washington University School of Medicine
Eduardo Perozo, Ph.D., University of Virginia School of Medicine
Wendy Suzuki, Ph.D., New York University

1997–1999

Ulrike I. Gaul, Ph.D., The Rockefeller University
Liqun Luo, Ph.D., Stanford University School of Medicine
Mark Mayford, Ph.D., University of California, San Diego
Peter Mombaerts, M.D., Ph.D., The Rockefeller University
Samuel L. Pfaff, Ph.D., Salk Institute for Biological Studies
David Van Vactor, Ph.D., Harvard Medical School

1996–1998

Paul W. Glimcher, Ph.D., New York University
Ali Hemmati-Brivanlou, Ph.D., The Rockefeller University
Donald C. Lo, Ph.D., Duke University Medical Center
Earl K. Miller, Ph.D., Massachusetts Institute of Technology
Tito A. Serafini, Ph.D., University of California, Berkeley
Jerry C.P. Yin, Ph.D., Cold Spring Harbor Laboratory

1995–1997

Toshinori Hoshi, Ph.D., University of Iowa
Alex L. Kolodkin, Ph.D., Johns Hopkins University School of Medicine

Michael L. Nonet, Ph.D., Washington University School of Medicine
Mani Ramaswami, Ph.D., University of Arizona
Michael N. Shadlen, M.D., Ph.D., University of Washington
Alcino J. Silva, Ph.D., Cold Spring Harbor Laboratory

1994–1996

Rita J. Balice-Gordon, Ph.D., University of Pennsylvania
Mark K. Bennett, Ph.D., University of California, Berkeley
David S. Bredt, M.D., Ph.D., University of California, San Francisco
David J. Linden, Ph.D., Johns Hopkins University School of Medicine
Richard D. Mooney, Ph.D., Duke University Medical Center
Charles J. Weitz, M.D., Ph.D., Harvard Medical School

1993–1995

Ben Barres, M.D., Ph.D., Stanford University School of Medicine
Allison J. Doupe, M.D., Ph.D., University of California, San Francisco
Ehud Y. Isacoff, Ph.D., University of California, Berkeley
Susan K. McConnell, Ph.D., Stanford University School of Medicine
John J. Ngai, Ph.D., University of California, Berkeley
Wade G. Regehr, Ph.D., Harvard Medical School

1992–1994

Ethan Bier, Ph.D., University of California, San Diego
Linda D. Buck, Ph.D., Harvard Medical School
Gian Garriga, Ph.D., University of California, Berkeley
Roderick MacKinnon, M.D., Harvard Medical School
Nipam H. Patel, Ph.D., Carnegie Institution of Washington
Gabriele V. Ronnett, M.D., Ph.D., Johns Hopkins University School of
    Medicine
Daniel Y. Ts'o, Ph.D., The Rockefeller University

1991–1993

Hollis T. Cline, Ph.D., University of Iowa Medical School
Gilles J. Laurent, Ph.D., California Institute of Technology

Ernest G. Peralta, Ph.D., Harvard University
William M. Roberts, Ph.D., University of Oregon
Thomas L. Schwarz, Ph.D., Stanford University School of Medicine
Marc T. Tessier-Lavigne, Ph.D., University of California, San Francisco

1990–1992

John R. Carlson, Ph.D., Yale University School of Medicine
Michael E. Greenberg, Ph.D., Harvard Medical School
David J. Julius, Ph.D., University of California, San Francisco
Robert C. Malenka, M.D., Ph.D., University of California, San Francisco
J. David Sweatt, Ph.D., Baylor College of Medicine
Kai Zinn, Ph.D., California Institute of Technology

1989–1991

Utpal Banerjee, Ph.D., University of California, Los Angeles
Paul Forscher, Ph.D., Yale University School of Medicine
Michael D. Mauk, Ph.D., University of Texas Medical School
Eric J. Nestler, M.D., Ph.D., Yale University School of Medicine
Barbara E. Ranscht, Ph.D., La Jolla Cancer Research Foundation

1988–1990

Michael Bastiani, Ph.D., University of Utah
Craig E. Jahr, Ph.D., Oregon Health & Science University
Christopher R. Kintner, Ph.D., Salk Institute for Biological Studies
Jonathan A. Raper, Ph.D., University of Pennsylvania Medical Center
Lorna W. Role, Ph.D., Columbia University College of Physicians and
    Surgeons
Charles Zuker, Ph.D., University of California, San Diego

1987–1989

Aaron P. Fox, Ph.D., University of Chicago
F. Rob Jackson, Ph.D., Worcester Foundation for Experimental Biology
Dennis D.M. O'Leary, Ph.D., Washington University School of Medicine
Tim Tully, Ph.D., Brandeis University
Patricia A. Walicke, M.D., Ph.D., University of California, San Diego

1986–1988

Christine E. Holt, Ph.D., University of California, San Diego
Stephen J. Peroutka, M.D., Ph.D., Stanford University School of Medicine
Randall N. Pittman, Ph.D., University of Pennsylvania School of Medicine
S. Lawrence Zipursky, Ph.D., University of California, Los Angeles

1985–1987

Sarah W. Bottjer, Ph.D., University of Southern California
S. Marc Breedlove, Ph.D., University of California, Berkeley
Jane Dodd, Ph.D., Columbia University College of Physicians and Surgeons
Haig S. Keshishian, Ph.D., Yale University School of Medicine
Paul E. Sawchenko, Ph.D., Salk Institute for Biological Studies

1984–1986

Ronald L. Davis, Ph.D., Baylor College of Medicine
Scott E. Fraser, Ph.D., University of California, Irvine
Michael R. Lerner, M.D., Ph.D., Yale University School of Medicine
William D. Matthew, Ph.D., Harvard Medical School
Jonathan D. Victor, M.D., Ph.D., Cornell University Medical College

1983–1985

Richard A. Andersen, Ph.D., Salk Institute for Biological Studies
Clifford B. Saper, M.D., Ph.D., Washington University School of Medicine
Richard H. Scheller, Ph.D., Stanford University School of Medicine
Mark Allen Tanouye, Ph.D., California Institute of Technology
George R. Uhl, M.D., Ph.D., Massachusetts General Hospital

1982–1984

Bradley E. Alger, Ph.D., University of Maryland School of Medicine
Ralph J. Greenspan, Ph.D., Princeton University
Thomas M. Jessell, Ph.D., Columbia University College of Physicians and
  Surgeons
Bruce H. Wainer, M.D., Ph.D., University of Chicago
Peter J. Whitehouse, M.D., Ph.D., Johns Hopkins University School of
  Medicine

1981–1983

David G. Amaral, Ph.D., Salk Institute for Biological Studies
Robert J. Bloch, Ph.D., University of Maryland School of Medicine
Stanley M. Goldin, Ph.D., Harvard Medical School
Stephen G. Lisberger, Ph.D., University of California, San Francisco
Lee L. Rubin, Ph.D., The Rockefeller University

1980–1982

Theodore W. Berger, Ph.D., University of Pittsburgh
Thomas H. Brown, Ph.D., City of Hope Research Institute
Steven J. Burden, Ph.D., Harvard Medical School
Corey S. Goodman, Ph.D., Stanford University School of Medicine
William A. Harris, Ph.D., University of California, San Diego

1978–1980

Robert P. Elde, Ph.D., University of Minnesota Medical School
Yuh-Nung Jan, Ph.D., Harvard Medical School
Eve Marder, Ph.D., Brandeis University
James A. Nathanson, M.D., Ph.D., Yale University School of Medicine
Louis F. Reichardt, Ph.D., University of California, San Francisco

1977–1979

Linda M. Hall, Ph.D., Massachusetts Institute of Technology
Charles A. Marotta, M.D., Ph.D., Harvard Medical School
Urs S. Rutishauser, Ph.D., The Rockefeller University
David C. Spray, Ph.D., Albert Einstein College of Medicine

McKNIGHT TECHNOLOGICAL INNOVATIONS IN NEUROSCIENCE AWARDS

2006–2007

Pamela M. England, Ph.D., University of California, San Francisco
Alan Jasanoff, Ph.D., Massachusetts Institute of Technology

Richard J. Krauzlis, Ph.D., and Edward M. Callaway, Ph.D., Salk Institute
  for Biological Studies
Markus Meister, Ph.D., Harvard University

2005–2006

Karl Deisseroth, M.D., Ph.D., Stanford University
Samie R. Jaffrey, M.D., Ph.D., Weill Medical College, Cornell University
Jeff W. Lichtman, M.D., Ph.D., Harvard University, and Kenneth Hayworth,
  University of Southern California
Alice Y. Ting, Ph.D., Massachusetts Institute of Technology

2004–2005

E.J. Chichilnisky, Ph.D., Salk Institute for Biological Studies, and A.M.
  Litke, Ph.D., Santa Cruz Institute for Particle Physics
Daniel T. Chiu, Ph.D., University of Washington
Susan L. Lindquist, Ph.D., Whitehead Institute for Biomedical Research
Daniel L. Minor, Jr., Ph.D., University of California, San Francisco
Stephen J. Smith, Ph.D., Stanford University School of Medicine

2003–2004

Stuart Firestein, Ph.D., Columbia University
David Heeger, Ph.D., New York University
Paul Slesinger, Ph.D., Salk Institute for Biological Studies

2002–2003

Liqun Luo, Ph.D., Stanford University
A. David Redish, Ph.D.; Babak Ziaie, Ph.D.; and Arthur G. Erdman, Ph.D.,
  University of Minnesota
Bernardo Sabatini, M.D., Ph.D., Harvard Medical School
Karel Svoboda, Ph.D., Cold Spring Harbor Laboratory

2001–2002

Helen M. Blau, Ph.D., Stanford University
Graham C.R. Ellis-Davies, Ph.D., MCP Hahnemann University

Dwayne Godwin, Ph.D., Wake Forest University School of Medicine
Seong-Gi Kim, Ph.D., University of Minnesota Medical School

2000–2001

Stephen Lippard, Ph.D., Massachusetts Institute of Technology
Partha Mitra, Ph.D., and Richard Andersen, Ph.D., California Institute of
    Technology
William Newsome, Ph.D., and Mark Schnitzer, Ph.D., Stanford University
    School of Medicine
Timothy Ryan, Ph.D., Weill Medical College of Cornell University, and
    Gero Miesenböck, Ph.D., Memorial Sloan-Kettering Cancer Center
Daniel Turnbull, Ph.D., New York University School of Medicine

1999–2000

Michael E. Greenberg, Ph.D., and Ricardo E. Dolmetsch, Ph.D., Boston
    Children's Hospital
Paul W. Glimcher, Ph.D., New York University
Leslie C. Griffith, M.D., Ph.D., and Jeffrey C. Hall, Ph.D., Brandeis
    University
Warren S. Warren, Ph.D., Princeton University

McKNIGHT NEUROSCIENCE OF BRAIN DISORDERS
AWARDS (ESTABLISHED AS THE McKNIGHT MEMORY AND
BRAIN DISORDERS AWARDS)

2006–2008

C. Michael Crowder, M.D., Ph.D., Washington University School of
    Medicine
Guoping Feng, Ph.D., Duke University Medical Center
Jill Morris, Ph.D., Northwestern University and Children's Memorial
    Research Center
Jeffrey Noebels, M.D., Ph.D., and Richard Gibbs, Ph.D., Baylor College of
    Medicine
Alexander Schier, Ph.D., Harvard University

2005–2007

Richard Andersen, Ph.D., California Institute of Technology
John Brigande, Ph.D., Oregon Health & Science University; and Stefan
  Heller, Ph.D., Harvard Medical School
William A. Catterall, Ph.D., University of Washington School of Medicine
Sacha B. Nelson, M.D., Ph.D., and Gina G. Turrigiano, Ph.D., Brandeis
  University
Gregory A. Petsko, Ph.D., and Dagmar Ringe, Ph.D., Brandeis University
Kai Zinn, Ph.D., California Institute of Technology

2004–2006

Seymour Benzer, Ph.D., California Institute of Technology
Valina L. Dawson, Ph.D., Johns Hopkins University School of Medicine
Alan L. Goldin, M.D., Ph.D., University of California, Irvine
Michael E. Greenberg, Ph.D., Children's Hospital and Harvard Medical
  School
Steven P. Hamilton, M.D., Ph.D., University of California, San Francisco
Irwin B. Levitan, Ph.D., University of Pennsylvania School of Medicine

2003–2005

Harvey Cantor, M.D., and Bruce Yankner, Ph.D., M.D., Harvard Medical
  School
Ronald L. Davis, Ph.D., Baylor College of Medicine
Joseph A. Gogos, M.D., Ph.D., Columbia University, College of Physicians
  & Surgeons
Paul H. Patterson, Ph.D., California Institute of Technology
Scott A. Small, M.D., Columbia University
Vivien Yee, Ph.D., The Cleveland Clinic Foundation
Gary I. Yellen, Ph.D., Harvard Medical School

2002–2004

Maureen Condic, Ph.D., University of Utah School of Medicine
Fen-Biao Gao, Ph.D., University of California, San Francisco
Kimberly Huber, Ph.D., University of Texas Southwestern Medical School
Pat Levitt, Ph.D., Vanderbilt University

Emmanuel Mignot, M.D., Ph.D., Stanford University Sleep Research Center
Christopher Walsh, M.D., Ph.D., Harvard Medical School and Beth Israel Deaconess Medical Center

2001–2003

Mel Feany, M.D., Ph.D., Harvard Medical School and Brigham and Women's Hospital
Oliver Hobert, Ph.D., Columbia University College of Physicians and Surgeons
Steven McKnight, Ph.D., University of Texas Southwestern Medical Center
Margaret Pericak-Vance, Ph.D., and Jeffery Vance, M.D., Ph.D., Duke University Medical Center
Stephen Strittmatter, M.D., Ph.D., Yale University School of Medicine

McKNIGHT INVESTIGATOR AWARDS (ESTABLISHED AS THE McKNIGHT DEVELOPMENT AWARDS; THESE AWARDS ARE NO LONGER OFFERED)

2000–2002

John R. Carlson, Ph.D., Yale University
Ulrike Heberlein, Ph.D., University of California, San Francisco
Alex L. Kolodkin, Ph.D., Johns Hopkins University School of Medicine
Gilles J. Laurent, Ph.D., California Institute of Technology
Dennis D.M. O'Leary, Ph.D., Salk Institute for Biological Studies

1997–1999

George J. Augustine, Ph.D., Duke University Medical Center
Utpal Banerjee, Ph.D., University of California, Los Angeles
Ben Barres, M.D., Ph.D., Stanford University School of Medicine
Stuart Firestein, Ph.D., Columbia University
David Fitzpatrick, Ph.D., Duke University Medical Center
David J. Julius, Ph.D., University of California, San Francisco
Roderick MacKinnon, M.D., The Rockefeller University
Robert C. Malenka, M.D., Ph.D., University of California, San Francisco
Gail Mandel, Ph.D., State University of New York

Thomas F.J. Martin, Ph.D., University of Wisconsin, Madison
Susan K. McConnell, Ph.D., Stanford University School of Medicine
David McCormick, Ph.D., Yale University School of Medicine
Jeffery D. Schall, Ph.D., Vanderbilt University
Terry T. Takahashi, Ph.D., University of Oregon
Gary Yellen, Ph.D., Massachusetts General Hospital

1994–1996

Susan G. Amara, Ph.D., Oregon Health & Science University
Bruce P. Bean, Ph.D., Harvard Medical School
Jane Dodd, Ph.D., Columbia University College of Physicians & Surgeons
Eric Frank, Ph.D., University of Pittsburgh School of Medicine
Philip G. Haydon, Ph.D., Iowa State University
Lawrence C. Katz, Ph.D., Duke University Medical Center
Chris R. Kintner, Ph.D., Salk Institute for Biological Studies
Nikos K. Logothetis, Ph.D., Baylor College of Medicine
Eve Marder, Ph.D., Brandeis University
Lawrence E. Mays, Ph.D., University of Alabama at Birmingham
Jonathan A. Raper, Ph.D., University of Pennsylvania Medical Center
Lorna W. Role, Ph.D., Research Foundation for Mental Hygiene
Marc T. Tessier-Lavigne, Ph.D., University of California, San Francisco
Kai Zinn, Ph.D., California Institute of Technology

1991–1993

Thomas D. Albright, Ph.D., Salk Institute for Biological Studies
Martin Chalfie, Ph.D., Columbia University
Gregor Eichele, Ph.D., Baylor College of Medicine
David Ferster, Ph.D., Northwestern University
Barry Ganetzky, Ph.D., University of Wisconsin
Charles Gilbert, M.D., Ph.D., The Rockefeller University
Michael R. Green, Ph.D., Harvard University
Mary E. Hatten, Ph.D., The Rockefeller University
John H.R. Maunsell, M.D., Ph.D., University of Rochester
Marc R. Montminy, Ph.D., Salk Institute for Biological Studies
Gary Ruvkun, Ph.D., Massachusetts General Hospital
Klaudiusz R. Weiss, Ph.D., Mount Sinai School of Medicine
Monte Westerfield, Ph.D., University of Oregon
S. Lawrence Zipursky, Ph.D., University of California, Los Angeles

1988–1990

Craig H. Bailey, Ph.D., Research Foundation for Mental Hygiene
Ronald L. Davis, Ph.D., Baylor College of Medicine
Kathleen Dunlap, Ph.D., Tufts University School of Medicine
Fred H. Gage, Ph.D., University of California, San Diego
William A. Harris, Ph.D., University of California, San Diego
Steven G. Lisberger, Ph.D., University of California, San Francisco
Kelly E. Mayo, Ph.D., Northwestern University
William T. Newsome, Ph.D., Stanford University School of Medicine
Nikolaos K. Robakis, Ph.D., The Mount Sinai Medical Center
Tsunao Saitoh, Ph.D., University of California, San Diego
Joshua R. Sanes, Ph.D., Washington University School of Medicine
Gary Struhl, Ph.D., College of Physicians & Surgeons, Columbia University
Mriganka Sur, Ph.D., Massachusetts Institute of Technology
Gary L. Westbrook, M.D., Oregon Health & Science University

1985–1987

Thomas H. Brown, Ph.D., Beckman Research Institute of the City of Hope
Corey S. Goodman, Ph.D., University of California, Berkeley
Mark Gurney, Ph.D., The University of Chicago
James F. Gusella, Ph.D., Harvard Medical School and Massachusetts
    General Hospital
Thomas M. Jessell, Ph.D., Columbia University College of Physicians &
    Surgeons
Mary B. Kennedy, Ph.D., California Institute of Technology
Eric I. Knudsen, Ph.D., Stanford University School of Medicine
Jeff W. Lichtman, M.D., Ph.D., Washington University School of Medicine
Charles Marotta, M.D., Ph.D., Massachusetts General Hospital
William G. Quinn, Ph.D., Massachusetts Institute of Technology
Samuel M. Schacher, Ph.D., Research Foundation for Mental Hygiene
Michael P. Stryker, Ph.D., University of California, San Francisco
Larry W. Swanson, Ph.D., Salk Institute for Biological Studies
Mark B. Willard, Ph.D., Washington University School of Medicine

1982–1984

Xandra O. Breakefield, Ph.D., Shriver Center for Mental Retardation
Gerald L. Hazelbauer, Ph.D., Washington State University

Daniel Johnston, Ph.D., Baylor College of Medicine
Jon M. Lindstrom, Ph.D., Salk Institute for Biological Studies
Richard E. Mains, Ph.D., Johns Hopkins University School of Medicine
Kenneth J. Muller, Ph.D., University of Miami School of Medicine
James A. Nathanson, M.D., Ph.D., Massachusetts General Hospital
Paul H. Patterson, Ph.D., California Institute of Technology
Carla J. Shatz, Ph.D., Stanford University School of Medicine
David C. Spray, Ph.D., Albert Einstein College of Medicine
James W. Truman, Ph.D., University of Washington
Thomas A. Woolsey, M.D., Washington University School of Medicine
Anne B. Young, M.D., Ph.D., University of Michigan

McKNIGHT SENIOR INVESTIGATOR AWARDS
(ESTABLISHED AS THE McKNIGHT AWARDS FOR RESEARCH
PROJECTS; THESE AWARDS ARE NO LONGER OFFERED)

1997–1999

William A. Catterall, Ph.D., University of Washington
Peter Dallos, Ph.D., Northwestern University
Norman Davidson, Ph.D., California Institute of Technology
Richard H. Goodman, M.D., Ph.D., Oregon Health & Science University
Eric I. Knudsen, Ph.D., Stanford University School of Medicine
Lynn T. Landmesser, Ph.D., Case Western Reserve University
Irwin B. Levitan, Ph.D., Brandeis University
Fernando Nottebohm, Ph.D., The Rockefeller University
Joshua R. Sanes, Ph.D., Washington University School of Medicine
Brian A. Wandell, Ph.D., Stanford University School of Medicine

1994–1996

Denis A. Baylor, M.D., Stanford University School of Medicine
Seymour Benzer, Ph.D., California Institute of Technology
William A. Harris, Ph.D., University of California, San Diego
Stephen F. Heinemann, Ph.D., Salk Institute for Biological Studies
Arthur Karlin, Ph.D., Columbia University College of Physicians and
    Surgeons
Story C. Landis, Ph.D., Case Western Reserve University
Ken Nakayama, Ph.D., Harvard University

Elio Raviola, M.D., Ph.D., Harvard Medical School
David L. Sparks, Ph.D., University of Pennsylvania Medical Center
Richard W. Tsien, Ph.D., Stanford University School of Medicine

1991–1993

Seymour Benzer, Ph.D., California Institute of Technology
Pietro De Camilli, M.D., Yale University School of Medicine
Claude P.J. Ghez, M.D., Research Foundation for Mental Hygiene
Stephen F. Heinemann, Ph.D., Salk Institute for Biological Studies
Bertil Hille, Ph.D., University of Washington School of Medicine
David H. Hubel, M.D., Harvard Medical School
Uel J. McMahan, II, Ph.D., Stanford University School of Medicine
Ricardo Miledi, M.D., University of California, Irvine
Larry R. Squire, Ph.D., University of California, San Diego
Nobuo Suga, Ph.D., Washington University School of Medicine

1988–1990

Ira B. Black, M.D., Cornell University Medical College
Joseph T. Coyle, M.D., Johns Hopkins University School of Medicine
Gerald D. Fischbach, M.D., Washington University School of Medicine
Bertil Hille, Ph.D., University of Washington Medical School
Masakazu Konishi, Ph.D., California Institute of Technology
Paul Patterson, Ph.D., California Institute of Technology
Michael G. Rosenfeld, M.D., University of California, San Diego
Dennis J. Selkoe, M.D., Brigham and Women's Hospital
Larry R. Squire, Ph.D., University of California, San Diego
Richard F. Thompson, Ph.D., University of Southern California

1985–1987

Ira B. Black, M.D., Cornell University Medical College
Joseph T. Coyle, M.D., Johns Hopkins University School of Medicine
James Douglass, Ph.D., Oregon Health & Science University
Gerald D. Fischbach, M.D., Washington University School of Medicine
Ann M. Graybiel, Ph.D., Massachusetts Institute of Technology
Marek-Marsel Mesulam, Ph.D., Beth Israel Hospital
Paul H. Patterson, Ph.D., California Institute of Technology

Michael G. Rosenfeld, M.D., University of California, San Diego
Charles F. Stevens, M.D., Ph.D., Yale University School of Medicine
Richard F. Thompson, Ph.D., University of Southern California

1984–1988

Robert D. Terry, M.D., and Peter Davies, Ph.D., Albert Einstein College of Medicine

1981–1983

Robert D. Terry, M.D., Albert Einstein College of Medicine

1980–1985

Samuel H. Barondes, M.D., University of California, San Francisco
Paul Greengard, Ph.D., The Rockefeller University
Eric Kandel, M.D., Columbia University College of Physicians and Surgeons
Fred Plum, M.D., Cornell University Medical College
Michael L. Shelanski, M.D., Ph.D., New York University Medical Center
Solomon H. Snyder, M.D., Johns Hopkins University School of Medicine

1977–1980

Samuel H. Barondes, M.D., University of California, San Diego
Paul Greengard, Ph.D., The Rockefeller University
Eric Kandel, M.D., Columbia University College of Physicians and Surgeons
James L. McGaugh, Ph.D., University of California, Irvine
Fred Plum, M.D., Cornell University Medical College
Victor E. Shashoua, Ph.D., Massachusetts General Hospital
Michael L. Shelanski, M.D., Ph.D., New York University Medical Center
Solomon H. Snyder, M.D., Johns Hopkins University School of Medicine
Thoralf M. Sundt, Jr., M.D., Mayo Medical School

Appendix III

By the Numbers: Major Awards to McKnight-Affiliated Scientists

NOBEL LAUREATES

McKnight Awardees

1. David Hubel, 1981 (Senior Investigator, 1991)
2. Eric Kandel, 2000 (Senior Investigator, 1977, 1980)
3. Paul Greengard, 2000 (Senior Investigator, 1977, 1980)
4. Roderick MacKinnon, 2003 (Scholar, 1992; Investigator, 1997)
5. Linda Buck, 2004 (Scholar, 1992)

Past Board Members and Advisers

1. Julius Axelrod, 1970
2. Torsten Wiesel, 1981

NATIONAL MEDAL OF SCIENCE RECIPIENTS

McKnight Awardees

1. Seymour Benzer, 1982 (Senior Investigator, 1991, 1994; Brain Disorders, 2004)
2. Eric Kandel, 1988 (Senior Investigator, 1977, 1980)
3. Ann Graybiel, 2001 (Senior Investigator, 1985)
4. Solomon Snyder, 2003 (Senior Investigator, 1977, 1980)
5. Stephen Lippard, 2004 (Technological Innovations, 2000)

NATIONAL ACADEMY OF SCIENCES MEMBERS

McKnight Awardees

1. Susan Amara (Investigator, 1994)
2. Richard Andersen (Scholar, 1983; Brain Disorders, 2005)
3. Denis Baylor (Senior Investigator, 1994)
4. Seymour Benzer (Senior Investigator, 1991, 1994; Brain Disorders, 2004)
5. Linda Buck (Scholar, 1992)
6. Harvey Cantor (Brain Disorders, 2003)
7. William Catterall (Senior Investigator, 1997; Brain Disorders, 2005)
8. Martin Chalfie (Investigator, 1991)
9. Gerald Fischbach (Senior Investigator, 1985, 1988)
10. Fred Gage (Investigator, 1988)
11. Barry Ganetzky (Investigator, 1991)
12. Charles Gilbert (Investigator, 1991)
13. Corey Goodman (Scholar, 1980; Investigator, 1985)
14. Richard Goodman (Senior Investigator, 1997)
15. Ann Graybiel (Senior Investigator, 1985)
16. Paul Greengard (Senior Investigator, 1977, 1980)
17. Jeffrey Hall (Technological Innovations, 1999)
18. Stephen Heinemann (Senior Investigator, 1991, 1994)
19. Bertil Hille (Senior Investigator, 1988, 1991)
20. David Hubel (Senior Investigator, 1991)
21. Thomas Jessell (Scholar, 1982; Investigator, 1985)
22. David Julius (Scholar, 1990; Investigator, 1997)
23. Eric Kandel (Senior Investigator, 1977, 1980)
24. Arthur Karlin (Senior Investigator, 1994)
25. Eric Knudsen (Investigator, 1985; Senior Investigator, 1997)
26. Masakazu Konishi (Senior Investigator, 1988)
27. Lynn Landmesser (Senior Investigator, 1997)
28. Susan Lindquist (Technological Innovations, 2004)
29. Stephen Lippard (Technological Innovations, 2000)
30. Roderick MacKinnon (Scholar, 1992; Investigator, 1997)
31. James McGaugh (Senior Investigator, 1977)
32. Steven McKnight (Brain Disorders, 2001)
33. Ricardo Miledi (Senior Investigator, 1991)
34. William Newsome (Investigator, 1988; Technological Innovations, 2000)
35. Fernando Nottebohm (Senior Investigator, 1997)
36. Gregory Petsko (Brain Disorders, 2005)

37. Michael Rosenfeld (Senior Investigator, 1985, 1988)
38. Joshua Sanes (Investigator, 1988; Senior Investigator, 1997)
39. Richard Scheller (Scholar, 1983)
40. Carla Shatz (Investigator, 1982)
41. Solomon Snyder (Senior Investigator, 1977, 1980)
42. Larry Squire (Senior Investigator, 1988, 1991)
43. Charles Stevens (Senior Investigator, 1985)
44. Nobuo Suga (Senior Investigator, 1991)
45. Marc Tessier-Lavigne (Scholar, 1991; Investigator, 1994)
46. Richard Thompson (Senior Investigator, 1985, 1988)
47. Richard Tsien (Senior Investigator, 1994)
48. Brian Wandell (Senior Investigator, 1997)
49. Charles Zuker (Scholar, 1988)

CURRENT AND PAST BOARD AND COMMITTEE MEMBERS AND ADVISERS

1. Julius Axelrod
2. Cori Bargman
3. Pietro De Camilli
4. Edward Evarts
5. Gerald Fischbach
6. Fred Gage
7. Charles Gilbert
8. Corey Goodman
9. Ann Graybiel
10. Paul Greengard
11. Stephen Heinemann
12. James Hudspeth
13. Thomas Jessell
14. David Julius
15. Eric Kandel
16. Seymour Kety
17. Robert Lefkowitz
18. Oliver Lowry
19. Jeremy Nathans
20. Eugene Roberts
21. Michael Rosenfeld
22. Richard Scheller
23. Carla Shatz
24. Solomon Snyder

25. Larry Squire
26. Charles Stevens
27. Lubert Stryer
28. David Tank
29. Richard Tsien
30. Roger Tsien
31. Torsten Wiesel
32. Robert Wurtz
33. Huda Zoghbi

HOWARD HUGHES MEDICAL INSTITUTE INVESTIGATORS (PAST AND PRESENT)

McKNIGHT AWARDEES

1. Thomas Albright (Investigator, 1991)
2. Susan Amara (Investigator, 1994)
3. Linda Buck (Scholar, 1992)
4. Pietro De Camilli (Senior Investigator, 1991)
5. Michael Ehlers (Scholar, 1999)
6. Corey Goodman (Scholar, 1980; Investigator, 1985)
7. Michael Green (Investigator, 1991)
8. Oliver Hobert (Brain Disorders, 2001)
9. Yuh-Nung Jan (Scholar, 1978)
10. Thomas Jessell (Scholar, 1982; Investigator, 1985)
11. Eric Kandel (Senior Investigator, 1977, 1980)
12. Lawrence Katz (Investigator, 1994)
13. Alex Kolodkin (Scholar, 1995; Investigator, 2000)
14. Michael Lerner (Scholar, 1984)
15. Susan Lindquist (Technological Innovations, 2004)
16. Stephen Lisberger (Scholar, 1981; Investigator, 1988)
17. Liqun Luo (Scholar, 1997; Technological Innovations, 2002)
18. Roderick MacKinnon (Scholar, 1992; Investigator, 1997)
19. Gail Mandel (Investigator, 1997)
20. John Maunsell (Investigator, 1991)
21. Steven McKnight (Brain Disorders, 2001)
22. Emmanuel Mignot (Brain Disorders, 2002)
23. William Newsome (Investigator, 1988; Technological Innovations, 2000)
24. Nipam Patel (Scholar, 1992)

25. Louis Reichardt (Scholar, 1978)
26. Fred Rieke (Scholar, 1999)
27. Michael Rosenfeld (Senior Investigator, 1985, 1988)
28. Richard Scheller (Scholar, 1983)
29. H. Sebastian Seung (Scholar, 2000)
30. Michael Shadlen (Scholar, 1995)
31. Carla Shatz (Investigator, 1982)
32. Stephen J. Smith (Technological Innovations, 2004)
33. Charles Stevens (Senior Investigator, 1985)
34. Gary Struhl (Investigator, 1988)
35. Karel Svoboda (Technological Innovations, 2002)
36. Larry Swanson (Investigator, 1985)
37. Marc Tessier-Lavigne (Scholar, 1991; Investigator, 1994)
38. George Uhl (Scholar, 1983)
39. Christopher A. Walsh (Brain Disorders, 2002)
40. Gary Yellen (Investigator, 1997; Brain Disorders, 2003)
41. Larry Zipursky (Scholar, 1986; Investigator, 1991)
42. Charles Zuker (Scholar, 1988)

CURRENT AND PAST BOARD AND COMMITTEE MEMBERS AND ADVISERS

1. Thomas Albright
2. David J. Anderson
3. Cori Bargman
4. Pietro De Camilli
5. Catherine Dulac
6. Corey Goodman
7. James Hudspeth
8. Thomas Jessell
9. Eric Kandel
10. Lawrence Katz
11. Robert Lefkowitz
12. John Maunsell
13. J. Anthony Movshon
14. Jeremy Nathans
15. Michael Rosenfeld
16. Michael Shadlen
17. Carla Shatz
18. Charles Stevens

19. Larry Swanson
20. Roger Tsien
21. Christopher A. Walsh
22. Larry Zipursky
23. Huda Zoghbi

Appendix IV

Index of Names

The following are cited in the book and/or have served the McKnight program as an adviser, board member, or committee member.

Abbott, Laurence
Brandeis University; Technological Innovations Committee, 2004–

Albright, Thomas
Salk Institute for Biological Studies; Investigator Award, 1991; Scholar
    Committee, 2001–2006

Andersen, Richard
California Institute of Technology; Scholar Award, 1983; Technological
    Innovations Award, 2000; Brain Disorders Award, 2004

Anderson, David J.
California Institute of Technology; Brain Disorders Committee, 2005–

Axelrod, Julius
1912–2004; 1970 Nobel laureate in medicine; chief of section on
    pharmacology at National Institute of Mental Health; early consultant to
    McKnight on neuroscience program; chaired Senior/Investigator
    Committee 1977–1986; committee member, 1986–1989; endowment fund
    director, 1986–1989; honorary director upon retirement

Bargmann, Cori
Rockefeller University; Scholar Committee, 2003–

Barondes, Samuel
Langley Porter Psychiatric Institute, University of California, San
    Francisco; Senior Investigator Award, 1977, 1980; endowment fund
    director, 1986–2002; president, 1990–1998; ex officio director, 2002–
    2003; member of 1976 ad hoc committee to create McKnight Foundation

program; member of original Scholars Committee; chair 1986–1989;
Brain Disorders Committee, 2001–2005

Barres, Ben, M.D.
Stanford University School of Medicine; Scholar Award, 1993; Investigator
Award, 1997

Bastiani, Michael
University of Utah; Scholar Award, 1988

Baylor, Denis
Stanford University School of Medicine (emeritus); Senior Investigator
Award, 1994

Benzer, Seymour
California Institute of Technology; Senior Investigator Award, 1991, 1994;
Brain Disorders Award, 2004

Binger, Erika
Great-granddaughter of William L. McKnight; daughter of James M. and
Patricia S. Binger; McKnight Foundation director, 1994–; chair, 2004–

Binger, James H. (Jim)
1916–2005; husband of Virginia McKnight Binger; McKnight Foundation
director and early proponent of the neuroscience program

Binger, James McKnight (Mac)
Grandson of William L. McKnight; son of Virginia and James Binger;
McKnight Foundation director, 1973–

Binger, Patricia (Pat)
Wife of James McKnight Binger; McKnight Foundation director, 1988–;
endowment fund director, 2005–

Binger, Virginia McKnight
1916–2002; daughter of William L. and Maude McKnight; president of The
McKnight Foundation, 1974–1987

Birk, Peggy J.
Interim president of The McKnight Foundation, August 2005–2006;
McKnight Foundation director, 2002–

Blau, Helen
Stanford University; Technological Innovations Award, 2001

Bottjer, Sarah
University of Southern California; Scholar Award, 1985

Boyle, Edwin, Jr.
1923–1978; cardiologist who created Charleston Heart Study in 1950s; later developed theory of improving older people's memories via hyperbaric pressure; William L. McKnight funded and participated in this research at Miami Heart Institute in the 1970s

Boynton, Cynthia
Granddaughter of William L. McKnight; daughter of Virginia and James Binger; McKnight Foundation director, 1974–; president, 1987–1999; endowment fund director, 1987–2004

Bredt, David
Vice president, Eli Lilly and company; Scholar Award, 1994

Brigande, John
Oregon Health & Science University; Brain Disorders Award, 2005

Buck, Linda
University of Washington; Nobel laureate in physiology or medicine, 2004; Scholar Award, 1992

Burden, Steven J.
New York University; Scholar Award, 1980

Burton, Charles
Minneapolis neurosurgeon who participated in Foundation-initiated review of Edwin Boyle's memory research in Miami in 1975

Carlson, John R.
Yale University; Investigator Award, 2000

Catterall, William
University of Washington; Senior Investigator Award, 1997; Brain Disorders Award, 2005

Chichilnisky, E.J.
Salk Institute for Biological Studies; Scholar Award, 2000; Technological Innovations Award, 2004

Cline, Hollis T.
Cold Spring Harbor Laboratory; Scholar Award, 1991

Cotzias, George
1918–1977; leading Parkinson's disease researcher at Brookhaven National Laboratories; early consultant to Fred Plum in developing McKnight neuroscience program

Dallos, Peter
Northwestern University; Senior Investigator Award, 1997

Davies, Peter
Albert Einstein College of Medicine; Senior Investigator Award, 1984

Davis, Ronald L.
Baylor College of Medicine; Scholar Award, 1984; Investigator Award,
    1988; Brain Disorders Award, 2003

Dawson, Valina L.
Johns Hopkins University; Brain Disorders Award, 2004

De Camilli, Pietro
Yale University; Senior Investigator Award, 1991; Scholar Committee, 2003–

DiAntonio, Aaron
Washington University; Scholar Award, 2002

Dodd, Jane
Columbia University College of Physicians & Surgeons; Scholar Award,
    1985; Investigator Award, 1994

Dolmetsch, Ricardo
Stanford University; Technological Innovations Award, 1999; Scholar
    Award, 2004

Doupe, Allison J.
University of California, San Francisco; Scholar Award, 1993; Scholar
    Committee, 2006–

Dulac, Catherine
Harvard University; Technological Innovations Committee, 1999–

Edelman, Gerald M.
Founder and director, the Neurosciences Institute in San Diego; Nobel
    laureate in medicine, 1972; adviser to Fred Plum on original proposal for
    McKnight neuroscience program in 1976

Eichele, Gregor
Baylor College of Medicine; Scholar Award, 1991

Evarts, Edward
1926–1985; headed Section on Physiology at National Institute of Mental
    Health; member of McKnight's ad hoc advisory committee, 1976; Senior/
    Investigator Awards Committee, 1977–1985

Ewald, Russell V.
1920–2000; executive vice president of The McKnight Foundation, 1974–1989; deeply involved with creation of neuroscience program; treasurer of endowment fund, 1986–1989

Feany, Mel
Harvard Medical School; Brain Disorders Award, 2001

Feldman, Daniel
University of California, San Diego; Scholar Award, 2001

Feller, Marla
University of California, San Diego; Scholar Award, 2002

Firestein, Stuart
Columbia University; Investigator Award, 1997; Technological Innovations Award, 2003

Fischbach, Gerald
Columbia University College of Physicians & Surgeons; Senior Investigator Award, 1985, 1988; endowment fund director, 1995–2000; Senior/Investigator Committee, 1989–2001; chair, 1996–2001

Fraser, Scott E.
California Institute of Technology; Scholar Award, 1984

Gage, Fred (Rusty)
Salk Institute for Biological Studies; Investigator Award, 1988; Brain Disorders Committee, 2001–2005

Gertler, Frank
Massachusetts Institute of Technology; Scholar Award, 2002

Gilbert, Charles D.
Rockefeller University; Investigator Award, 1991; Brain Disorders Committee, 2001–

Glimcher, Paul W.
New York University; Scholar Award, 1996; Technological Innovations Award, 1999

Godwin, Dwayne
Wake Forest University; Technological Innovations Award, 2001

Goldberg, Nelson
1931–1999; professor of biochemistry at University of Minnesota; member of original Scholar Award Committee, 1977–1986

Goodman, Corey S.
Co-founder, president, and CEO, Renovis, Inc.; and adjunct professor, University of California, Berkeley; Scholar Award, 1980; endowment fund director, 1986–; president, 2000–2005; Scholar Committee, 1986–1997; chair, 1990–1997

Graybiel, Ann
Massachusetts Institute of Technology; Senior Investigator Award, 1985; endowment fund director, 1986–1996; Senior/Investigator Committee, 1989–1996

Greenberg, Michael E.
Harvard Medical School; Scholar Award, 1990; Technological Innovations Award, 1999; Brain Disorders Award, 2004

Greengard, Paul
Rockefeller University; Nobel laureate in physiology or medicine, 2000; Senior Investigator Award, 1977, 1980; Scholar Committee, 1986–1990

Gurney, Mark
Senior vice president, deCODE Genetics; Investigator Award, 1985

Gusella, James F.
Harvard Medical School; Investigator Award, 1985

Harris, William A.
University of Cambridge; Scholar Award, 1980; Investigator Award, 1988; Senior Investigator Award, 1994

Hatten, Mary E.
Rockefeller University; Investigator Award, 1991; Scholar Committee, 1994–2000

Heberlein, Ulrike
University of California, San Francisco; Investigator Award, 2000

Heeger, David
New York University; Technological Innovations Award, 2003

Heinemann, Stephen F.
Salk Institute for Biological Studies; Senior Investigator Award, 1991, 1994; Senior/Investigator Committee, 1996–1999; Technological Innovations Committee, 2003–

Heller, Stefan
Stanford University; Brain Disorders Award, 2005

Hille, Bertil
University of Washington; Senior Investigator Award, 1988, 1991

Holt, Christine
University of Cambridge; Scholar Award, 1986

Hubel, David H.
Harvard University (emeritus); Nobel laureate in physiology or medicine, 1981; Senior Investigator Award, 1991

Huber, Kimberly
University of Texas Southwestern Medical Center; Brain Disorders Award, 2002

Hudspeth, James
The Rockefeller University; Scholar Committee, 2001–2004

Isaacson, Jeffry
University of California, San Diego; Scholar Award, 2000

Jagadeesh, Bharathi
University of Washington; Scholar Award, 2002

Jahr, Craig
Oregon Health & Science University; Scholar Award, 1988; Scholar Committee, 2005–

Jarvik, Murray
University of California, Los Angeles; Senior/Investigator Committee, 1977–1984

Jessell, Thomas M.
Columbia University College of Physicians & Surgeons; Scholar Award, 1982; Investigator Award, 1985; endowment fund director, 1998–; Scholar Committee, 1986–; chair, 1998–

Johnston, Daniel
University of Texas; Investigator Award, 1982

Julius, David J.
University of California, San Francisco; Scholar Award, 1990; Investigator Award, 1997; endowment fund director, 2004–; Technological Innovations Committee, 2005–

Kandel, Eric
Columbia University College of Physicians & Surgeons; Nobel laureate in physiology or medicine, 2000; Senior Investigator Award, 1977, 1980; endowment fund director, 1986–1998; Senior/Investigator Committee, 1984–1995; chair, 1987–1995

Karlin, Arthur
Columbia University; Discretionary Award, 1986; Senior Investigator
    Award, 1994

Katz, Lawrence C. (Larry)
1956–2005; professor, Duke University Medical Center; Investigator Award,
    1994; Scholar Committee, 1996–2001; Technological Innovations
    Committee, 2002–2005

Kety, Seymour
1915–2000; chief, Laboratory of Clinical Science, National Institute of
    Mental Health; member of ad hoc advisory committee to new McKnight
    program, 1976; Senior/Investigator Committee, 1977–1986; endowment
    fund director, 1986–1989

Kim, Seong-Gi
University of Pittsburgh; Technological Innovations Award, 2001

King, Fred
University of Florida and Emory University (retired); member of the team
    invited to review Edwin Boyle's research in Miami in 1975

Knudsen, Eric I.
Stanford University; Investigator Award, 1985; Senior Investigator Award,
    1997

Kolodkin, Alex
Johns Hopkins University; Scholar Award, 1995; Investigator Award, 2000

Konishi, Masakazu
California Institute of Technology; Senior Investigator Award, 1988

Krauzlis, Richard
Salk Institute for Biological Studies; Scholar Award, 2000; Technological
    Innovations Award, 2006

Landmesser, Lynn
Case Western Reserve University; Senior Investigator Award, 1997

Laurent, Gilles J.
California Institute of Technology; Scholar Award, 1991; Investigator
    Award, 2000

Lefkowitz, Robert
Duke University Medical Center; Senior/Investigator Committee,
    1987–1990

Lerner, Michael R.
Arena Pharmaceuticals; Scholar Award, 1984

Levitan, Irwin
University of Pennsylvania School of Medicine; Senior Investigator Award,
1997; Brain Disorders Award, 2004

Levitt, Pat
Vanderbilt University; Brain Disorders Award, 2002

Lichtman, Jeff W.
Harvard University; Investigator Award, 1985; Technological Innovations
Award, 2005

Lippard, Stephen
Massachusetts Institute of Technology; Technological Innovations Award,
2000

Lisberger, Stephen
University of California, San Francisco; Scholar Award, 1981; Investigator
Award, 1988

Logothetis, Nikos
Max Planck Institute; Investigator Award, 1994

Lowry, Oliver
1910–1996; Washington University; Senior/Investigator Committee,
1976–1986

Luo, Liqun
Stanford University; Scholar Award, 1997; Technological Innovations
Award, 2002

Malenka, Robert
Stanford University; Scholar Award, 1990; Investigator Award, 1997

Martin, Kelsey
University of California, Los Angeles; Scholar Award, 2001

Mauk, Michael
University of Texas, Houston, Medical School; Scholar Award, 1989

Maunsell, John
Baylor College of Medicine; Senior/Investigator Committee, 1998–1999

Mayford, Mark
Scripps Research Institute; Scholar Award, 1997

McConnell, Susan
Stanford University; Scholar Award, 1993; Investigator Award, 1997

McGaugh, James
University of California, Irvine; member of ad hoc advisory committee to
    original McKnight program, 1976; Senior Investigator Award, 1977;
    Scholar Committee, 1977–1985

McKnight, Evelyn Franks
1914–1999; second wife of William L. McKnight; founder of McKnight
    Brain Research Foundation

McKnight, Maude
1882–1973; first wife of William L. McKnight and first vice president of
    The McKnight Foundation

McKnight, William L.
1887–1978; former president and CEO of the 3M Company; founder of
    The McKnight Foundation; his deep interest in learning and memory led
    to the foundation's neuroscience program

McMahan, Uel J.
Stanford University; Senior Investigator Award, 1991

Meister, Markus
Harvard University; Technological Innovations Award, 2006; Scholar
    Committee, 2003–2006

Mesulam, Marek-Marsel
Northwestern University; Senior Investigator Award, 1985

Mignot, Emmanuel
Stanford University; Brain Disorders Award, 2002

Miller, Earl K.
Massachusetts Institute of Technology; Scholar Award, 1996

Minor, Daniel L.
University of California, San Francisco; Scholar Award, 2001;
    Technological Innovations Award, 2004

Mitra, Partha
Cold Spring Harbor Laboratory; Technological Innovations Award, 2000

Mombaerts, Peter
Rockefeller University; Scholar Award, 1997

Mooney, Richard D.
Duke University Medical School; Scholar Award, 1994

Movshon, J. Anthony
New York University; Technological Innovations Committee, 1999–2004

Nathans, Jeremy
Johns Hopkins University School of Medicine; Brain Disorders Committee, 2001–

Nestler, Eric
University of Texas Southwestern Medical School; Scholar Award, 1989; endowment fund director, 2002–; Brain Disorders Committee, 2001–

Newsome, William
Stanford University; Investigator Award, 1988; Technological Innovations Award, 2000

Ngai, John
University of California, Berkeley; Scholar Award, 1993

Nottebohm, Fernando
Rockefeller University; Senior Investigator Award, 1997

O'Keefe, Michael
Executive vice president of The McKnight Foundation 1989–1999; president of the Minneapolis College of Art & Design, 2002–

O'Leary, Dennis D.M.
Salk Institute for Biological Studies; Scholar Award, 1987; Investigator Award, 2000

Patel, Nipam H.
University of California, Berkeley; Scholar Award, 1992

Patterson, Paul H.
California Institute of Technology; Investigator Award, 1982; Senior Investigator Award, 1985, 1988; Brain Disorders Award, 2003

Petsko, Gregory
Brandeis University; Brain Disorders Award, 2005

Pfaff, Samuel
Salk Institute for Biological Studies; Scholar Award, 1997

Pidany, Marilyn
Vice president for administration at The McKnight Foundation, 1976–1999; handled administrative matters for The McKnight Endowment Fund; now retired

Plum, Fred
Cornell Medical School (emeritus); consultant who helped The McKnight
    Foundation develop its original neuroscience program in 1975; Senior
    Investigator Award, 1977, 1980; Scholar Committee, 1977–1986;
    endowment fund director, 1986–1989; president, 1986–1989

Ranscht, Barbara
Burnham Institute; Scholar Award, 1989

Raper, Jonathan
University of Pennsylvania; Scholar Award, 1988; Investigator Award, 1994

Rapson, Rip
President of The McKnight Foundation, 1999–2005; president of the Kresge
    Foundation, 2006–

Ray, Charles
Minneapolis spine surgeon who was on the team invited by The McKnight
    Foundation to review Edwin Boyle's memory research in Miami
    in 1975

Redish, A. David
University of Minnesota; Technological Innovations Award, 2002

Reichardt, Louis
University of California, San Francisco; Scholar Award, 1978

Ringe, Dagmar
Brandeis University; Brain Disorders Award, 2005

Robakis, Nikolaos
Mount Sinai School of Medicine; Investigator Award, 1988

Roberts, Eugene
City of Hope Beckman Research Institute; Senior/Investigator Committee,
    1977–1984

Roelink, Henk
University of Washington; Scholar Award, 1999

Role, Lorna W.
Columbia University; Scholar Award, 1988; Investigator Award, 1994

Ronnett, Gabriele
Johns Hopkins University; Scholar Award, 1992

Rosenfeld, Michael G.
University of California, San Diego; Senior Investigator Award, 1985, 1988;
    Senior/Investigator Committee, 1990–1995

Rutishauser, Urs
Memorial Sloan-Kettering Cancer Center; Scholar Award, 1977

Ruvkun, Gary
Harvard Medical School; Investigator Award, 1991

Rysted, Kathleen
Director of research programs at The McKnight Foundation; oversees The
    McKnight Endowment Fund for Neuroscience

Sanes, Joshua R.
Harvard University; Investigator Award, 1988; Senior Investigator Award,
    1997

Saper, Clifford B.
Harvard Medical School; Scholar Award, 1983

Scheller, Richard H.
Executive vice president, Genentech; Scholar Award, 1983; Senior/
    Investigator Committee, 1995–1999

Schneckloth, Roland E.
As professor at Cornell University Medical College in 1976, advised Fred
    Plum on creation of original McKnight neuroscience program

Selkoe, Dennis
Harvard Medical School; Senior Investigator Award, 1988

Serafini, Tito
Vice president and co-founder, Renovis; Scholar Award, 1996

Shadlen, Michael N.
University of Washington; Scholar Award, 1995; Technological Innovations
    Committee, 2004–

Shatz, Carla J.
Harvard Medical School; Investigator Award, 1982; endowment fund
    director, 1998–; president, 2005–; Scholar Committee, 1991–1993;
    Senior/Investigator Committee, 1998–1999; Brain Disorders Committee,
    2001–2005

Shelanski, Michael L.
Columbia University Medical Center; Senior Investigator Award, 1977, 1980

Shepherd, John
Prominent cardiologist at Mayo Clinic who proposed a McKnight brain
    research laboratory in 1977

Silva, Alcino
University of California, Los Angeles; Scholar Award, 1995

Slesinger, Paul A.
Salk Institute for Biological Studies; Scholar Award, 1999; Technological
    Innovations Award, 2003

Small, Scott A.
Columbia University; Brain Disorders Award, 2003

Smith, Stephen J.
Stanford University; Technological Innovations Award, 2004

Snyder, Solomon
Johns Hopkins University; Scholar Committee, 1977–1993

Squire, Larry
University of California, San Diego; endowment fund director, 1999–;
    Scholar Committee, 1995–2000; Brain Disorders Committee, 2001–;
    chair, 2001–

Staryk, Noa
Great-granddaughter of William L. McKnight; daughter of Cynthia Binger
    Boynton and her former husband, George Boynton; chair of The
    McKnight Foundation, 1999–2004; director 1991–2005; married to Ted
    Staryk, a McKnight Foundation director

Stevens, Charles F. (Chuck)
Salk Institute for Biological Studies; Senior Investigator Award, 1985;
    endowment fund director, 1986–1996; Scholar Committee, 1986–1996

Stryer, Lubert
Stanford University; endowment fund director, 1998–; Technological
    Innovations Committee, 1999–2004; chair, 1999–2004

Stryker, Michael
University of California, San Francisco; Investigator Award, 1985; Scholar
    Committee, 1992–1999

Sur, Mriganka
Massachusetts Institute of Technology; Investigator Award, 1988

Swanson, Larry W.
University of Southern California; Investigator Award, 1985; Scholar
    Committee, 1986–1995

Sweatt, J. David
University of Alabama, Birmingham; Scholar Award, 1990

Tank, David
Princeton University; endowment fund director, 2005–; Technological
  Innovations Committee, 1999–; chair, 2005–

Terry, Robert D.
University of California, San Diego (emeritus); Senior Investigator Award,
  1984, 1987; Senior/Investigator Committee, 1986–1989

Tessier-Lavigne, Marc
Senior vice president, Genentech; Scholar Award, 1991; Investigator Award,
  1994

Thomas, Lewis
1913–1993; CEO of Memorial Sloan-Kettering Cancer Institute; author;
  adviser to Fred Plum on McKnight neuroscience proposal, 1976

Thompson, Richard F.
University of Southern California; Senior Investigator Award, 1985, 1988

Tower, Donald
Former director of the National Institute of Neurological Disorders and
  Stroke; adviser to Fred Plum on McKnight neuroscience proposal, 1976

Tsien, Richard
Stanford University; Scholar Committee, 1998–2003

Tsien, Roger
University of California, San Diego; Technological Innovations Committee,
  1999–2002

Tully, Tim
Cold Spring Harbor Laboratory; Scholar Award, 1987

Turnbull, Daniel
New York University School of Medicine; Technological Innovations
  Award, 2000

Van Vactor, David
Harvard Medical School; Scholar Award, 1997

Victor, Jonathan D.
Weill Medical College of Cornell University; Scholar Award, 1984

Vosshall, Leslie
Rockefeller University; Scholar Award, 2001

Wagner, Anthony D.
Stanford University; Scholar Award, 2001

Walsh, Chris A.
Harvard Medical School; Brain Disorders Award, 2002; Brain Disorders
   Committee, 2005–

Wandell, Brian
Stanford University; Senior Investigator Award, 1997

Watson, James
Nobel laureate in 1962 for co-discovery of double helix structure of DNA;
   Director's Award, 1984

Weliky, Michael
University of Rochester; Scholar Award, 1999

Westbrook, Gary L.
Oregon Health & Science University; Investigator Award, 1988

Westerfield, Monte
University of Oregon; Investigator Award, 1991

Whitehouse, Peter J.
Case Western Reserve University; Scholar Award, 1982

Wiesel, Torsten
Rockefeller University (president emeritus); Nobel laureate in physiology or
   medicine, 1981; endowment fund director, 1991–2003; president, 1999–
   2000; Senior/Investigator Committee, 1990–1999

Wurtz, Robert
Section chief, National Eye Institute, who worked with Edward Evarts at
   NIH; Senior/Investigator Committee, 1986–1998; Technological
   Innovations Committee, 1999–2003

Yang, Jian
Columbia University; Scholar Award, 2000

Yellen, Gary
Harvard Medical School; Investigator Award, 1997; Brain Disorders Award,
   2003

Yin, Jerry C.P.
University of Wisconsin; Scholar Award, 1996

Young, Anne B.
Massachusetts General Hospital; Investigator Award, 1982

Zinn, Kai
California Institute of Technology; Scholar Award, 1990; Investigator
   Award, 1994; Brain Disorders Award, 2005

Printed and bound by CPI Group (UK) Ltd, Croydon, CR0 4YY

03/10/2024

01040412-0016

Zipursky, S. Lawrence (Larry)
University of California, Los Angeles; Scholar Award, 1986; Investigator
Award, 1991; Scholar Committee, 1997–2004

Zoghbi, Huda Yahya
Baylor College of Medicine; endowment fund director, 2003–; Brain
Disorders Committee, 2005–

Zuker, Charles S.
University of California, San Diego; Scholar Award, 1988

Index

Page numbers with "f" denote figures